Liz Jones

About the Author

Dr. Robert Goodwin is a visiting research fellow at University College London. He has also studied at King's College London, the School of Oriental and African Studies, and in Spain at the universities of Granada and Seville. He lives in London.

Los "mulatos" de Esmeraldas, by Adrián Sánchez Galque (c. 1599). (*Courtesy of Museo de América, Madrid, Spain*)

CROSSING THE CONTINENT 1527–1540

The Story of the First African-American Explorer of the American South

ROBERT GOODWIN

HARPER

NEW YORK ▪ LONDON ▪ TORONTO ▪ SYDNEY

HARPER

FIRST HARPER PAPERBACK PUBLISHED 2009.

Designed by Level C

Library of Congress Cataloging-in-Publication Data
is available upon request.

ISBN 978-0-06-114045-7

09 10 11 12 13   WBC/RRD   10 9 8 7 6 5 4 3 2 1

For Liz

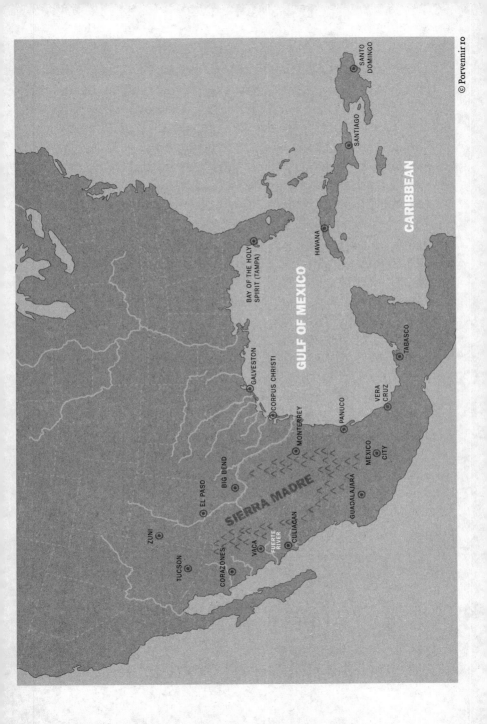

SANTO DOMINGO

SANTIAGO

CARIBBEAN

HAVANA

BAY OF THE HOLY
SPIRIT (TAMPA)

GULF OF MEXICO

GALVESTON

CORPUS CHRISTI

TABASCO

PANUCO

VERA
CRUZ

MONTERREY

MEXICO
CITY

BIG BEND

EL PASO

SIERRA MADRE

GUADALAJARA

ZUNI

VACA

CULIACAN

FUERTE
RIVER

TUCSON

CORAZONES

© Porvenni rio

CONTENTS

Part Two
ESTEBAN

Part Three
WRITING HISTORY

Part Four
THE EXPEDITION TO FLORIDA
1527–1536

LIST OF ILLUSTRATIONS

ACKNOWLEDGMENTS

ESPECIAL THANKS GO to Barry Ife, a good friend and an inspiring teacher with a terrifying work ethic, and to Trudi Darby for all her support. The British Academy supported some of my research in Seville out of which the idea for this book was born. Paul Hoffman offered timely encouragement. The staff of the Archivo General de Indias makes it an exceptional place to work, y se lo agradezco especialmente a Pilar Lázaro de la Escosura la búsqueda del Santo Domingo 11 43 bis.

I must thank Alicia Ríos for introducing me to Michael Jacobs, to whom I am deeply indebted for years of encouragement, for introducing me to George Miller, who introduced me to my agent Rebecca Winfield, who is a credit to a much maligned profession. Louise, Arthur, and Elizabeth Jones read early drafts of the manuscript and made very helpful suggestions, while Claire Wachtel's erudite and critically incisive editing was a pleasure to work with. Thanks to everyone else at HarperCollins in New York, especially Julia Novitch for fielding my various grumbles. It was essential that Romolo, Claudio, and Tom kept me fed. Thanks to Nick for his company on all those Mexican buses and trains.

En Sevilla, Ana María Rengel, Manu, Javi, Bea, Cristina, Manuel, y Ana me recibieron como otro más de la familia y me ofrecieron el Porvenir como mi propio hogar sevillano. Gracias. También me presentaron a Jim, Jana, Emilio, y Andrea Anaya, mis cariñosos anfitriones en Tucson, ¡Viva el chihuahua! En Madrid, Cuqui y Jesús me han recibido con los brazos y las puertas abiertos. Gracias por todo a David, Carolina, Pedro y Isa, Anchy y Miguel, Lupe, Loly, Jesús, y las Glorias, los Luises, Carmen, Pilar, Falique y Teresa, los Perales y Pepe.

My journey to Zuni with Jim Anaya and Rob Williams was unforgettable. Ed Wemytewa and Jim Enote were amazingly kind and forthcoming, and Tom Kennedy was a gracious host. Sonahchi? Thanks to everyone at Zuni, especially the basketball team. Good luck. Gracias a Miguel por los elotes, a Wilfredo por la aventura del Río Fuerte, y a Omar y Odette en Oaxaca.

And a very special thank you to Jim Read, who first drew my attention to Cabeza de Vaca's account and the story of the Narváez expedition.

It remains only to thank Rolena Adorno and Charles Patrick Pautz for their remarkable *Alvar Núñez Cabeza de Vaca,* which is the bible or encyclopaedia for scholars working on Narváez's expedition to Florida.

MAIN CHARACTERS

Note: Each person is generally referred to by the italicized portion of his name.

Charles V, Holy Roman Emperor Charles V, Charles I of Castillo, the most powerful man in the world.

Pánfilo *Narváez*, Captain General and Governor of Florida.

Esteban, an African slave.

Andrés *Dorantes* de Carranza, Esteban's owner, a captain on the expedition to Florida.

Alvar Núñez *Cabeza de Vaca*, treasurer of the expedition to Florida, author of *Shipwrecks*.

Alonso del *Castillo* Maldonado, a captain on the expedition to Florida, a doctor's son.

Nuño Beltrán de *Guzmán*, an "evil" conquistador.

Diego de *Alcaraz* and Lázaro de *Cebreros*, Guzmán's henchmen.

Melchior *Díaz*, captain at Culiacán.

Antonio de *Mendoza*, Viceroy of Mexico.

Hernán *Cortés*, conqueror of Mexico.

Juan de *Zumárraga*, Archbishop and Inquisitor of Mexico.

Juan *Garrido,* a free African, resident of Mexico City.

Friar *Marcos* **de Niza,** a Franciscan monk, religious leader of the expedition to the Seven Cities of Gold.

Francisco Vázquez de *Coronado,* Mendoza's right-hand man.

Gonzalo Fernández de *Oviedo* **y Valdés,** Charles V's Historian Royal.

CHRONOLOGY

—◆—

c. 1500 Esteban is born in Africa.

c. 1522 Esteban arrives in Spain.

1527 Narváez's expedition sails from Spain for the Caribbean.

1528 Easter: expedition lands near Tampa Bay.

1528 Fall: survivors land on *Malhado*, probably Galveston Island, Texas; disease strikes.

A handful of survivors live with the Karankawa Indians of the Texas coast.

1533 Only four men left alive: Esteban, Dorantes, Castillo, and Cabeza de Vaca.

1534 The four leave the Karankawa and head inland.

1535 The four travel through northeast Mexico, go up the Rio Grande, turn west at El Paso, spend Christmas at Corazones.

1536 March: the four meet Spanish slavers in northwest Mexico, reach Mexico City.

July: the four make their official report in Mexico City.

1537   February: Cabeza de Vaca and Dorantes try to sail for Spain.

August: Cabeza de Vaca lands at Lisbon and goes to the Spanish court.

November–December: Dorantes returns to Mexico after months lost at sea.

1538   Winter: Dorantes appointed as captain of cavalry for the expedition to the Seven Cities.

Summer: Dorantes is sidelined; Esteban is appointed as a "guide" and military leader of the expedition to the Seven Cities; Marcos de Niza is appointed as religious leader.

Fall: expedition to the Seven Cities leaves Mexico.

1539   Easter: Esteban leaves Marcos at Vacapa in northeastern Mexico.

Spring: Esteban reaches Zuni Pueblos, known as the Seven Cities of Cibola.

Summer–fall: Marcos returns to Mexico City reporting Esteban's death.

Fall: Mendoza organizes a major expedition to the Seven Cities, commanded by Coronado.

1540   Coronado reaches Zuni Pueblos; further reports of Esteban's death

INTRODUCTION

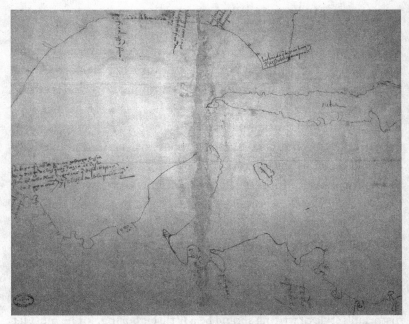

The earliest known map of the Gulf of Mexico (c. 1519). The River and Bay of the Holy Spirit ("Río del Espíritu Santo") is the last feature labeled toward the west of the north coast of the gulf. (*Courtesy of Archivo General de Indias; MP México 5*)

"Look at Odysseus's journey or Jason's voyage or the labors of Hercules, they are but fiction and fable. So read them as such for that is how they should be read. And do not admire the wonders in them, for they bear no comparison with the hardships of these sinners, who traveled such an unhappy road."

(Fernández de Oviedo, Spanish Historian Royal, c. 1540)

THIS IS THE story of how history is written, the history of Esteban's story, and also the tale of the first men in history to cross North America. It is a narrative of uncertainty, conjecture, and historical truth.

HISTORY IS THE origin myth of the white man. It tells us about our ancestors, their heroes and wars, about how we came to live as we do, about our gods and our morality. It defines our values and reveres our political institutions. It offers a continuous story of our civilization, from ancient Greece and Rome right up to the foundation of our own nation-states. I write, evidently enough, from the perspective of my own personal history.

We were taught from childhood that history is the true story of the past, based on facts. We learned that it is not the historian's job to dramatize his story in order to make it exciting. We learned that the historian's style of writing should be a little bit boring, for he must let the facts speak for themselves. It is his job to tell the truth, only the truth, and as much of the truth as he can. As the great Spanish novelist Miguel de Cervantes explained in 1615:

A poet may speak or sing of events, not as they really happened, but as they should have been; but the historian must write down events, not as they should have been, but as they really happened, neither embellishing nor suppressing anything that is true.

And yet we might pause to question if that is possible. How do we know that history is fact? How does a historian know that what he tells us is true?

As we all learned, as soon as our parents dared to let us know such things, any story we are told may be true or false. We learned quickly to talk of "tall stories" with a hint of smile, and we even praise a "good story," with something of a wink in our eye. These are the tales traditionally told by hunters, fishermen, soldiers, and other travelers. These are the stories that are based on facts but which are embellished with fiction. Such stories are entertaining, but more often than not they are also the boasts of a hero who tells his own tale. For which reason, we do not believe him. But do we believe him if the heroes of his stories are his parents, or his grandparents?

What should we make of the legend of King Arthur and the Knights of the Round Table? Clearly, it is not factual history, with its magical swords and invincible heroes, but what kind of story is it? How many of us are bold enough to say it is not true? How different is what we know about King Arthur from what we know of George Washington or Billy the Kid?

In recent years, many historians have been intrigued by this new manifesto of self-doubt, while others have exploded with apoplexy at this revolution from within their subject. In order to understand the history of how historians have arrived at this collective sense of uncertainty, it is useful to look briefly at some of the different ways in which the history of the Spanish conquest of Mexico has been written——not least because to do so also helps to explain the background to the story I am going to tell in this book.

Probably the most influential work written in English about this subject is *The History of the Conquest of Mexico* by William Hickling Prescott (1796–1859). As a law student in Boston, he was blinded by a youthful prank and was forced to give up the law. Fortunately, he instead turned his attention to writing history. He is said to have had such a prodigious memory that he was able to compose the chapters for his books while out riding in the morning and then write them up in the afternoon, using a special writing contraption for the blind which he had bought in London. His writing style suggests that he had a formidable personality, and he wrote history with a strong sense of drama and great literary flair. What he saw in his mind's eye more than made up for the real world he could not see.

Prescott was perhaps the most brilliant historian of his age, but he was also very much a child of his time and of the patrician social class into which he was born. As a result, his account is an aristocratic drama about his central character, Hernán Cortés, the commander of the first European army to march into Mexico and reach the Aztec capital, in 1519. Prescott used the work of many Spanish historians who had glorified Cortés, but he set particular store by the letters that Cortés himself sent back to Spain describing his discoveries and conquests. Yet was Prescott right to believe what Cortés had written? Did he stop to wonder critically enough why Cortés wrote them? Did a Spanish general have any reason to tell the truth?

Yes and no. The basics are certainly true. A small force of Spaniards, perhaps as many as 1,500, seized control of the mighty Aztec Empire. But Cortés was also a turncoat and a rebel who had betrayed his own superior, the governor of Cuba appointed by the Spanish crown. Cortés wrote his letters in order to prove that he was no traitor to his sovereign, the Holy Roman Emperor, Charles V, king of Spain, to convince Charles that he done nothing wrong and was in fact an imperial hero. But he also wrote in order to dazzle the Spanish court with a brilliant story of bravery and fabulous wealth, and he took care to send those letters to Spain along with tangible proof of his success in the form of gold and silver.

Cortés's letters tell a story that is ideal material for a man like Prescott with a strong sense of republican patrician honor. An inspiring and rebellious general drawn from the gentry led a small army of intrepid soldiers to an astonishing and glorious victory won against all odds over a great empire. They founded a new "republic," full of hope. God had clearly been on their side. As a parable, it suited Prescott and it suited America. But is it too good to be true? After all, how did a handful of Spaniards conquer the great Aztec Empire? It sounds more like an Arthurian legend than fact.

As the western world became more democratic, explanations for the conquest of Mexico began to embrace the humble as well as the mighty. The history of Mexico came to be seen through the eyes of one of the less important soldiers in Cortés's army, Bernal Díaz del Castillo. Díaz wrote his *True History of the Conquest of New Spain* in the 1550s because he was outraged that Cortés had contrived to claim all the glory for himself in the official account. But the *True History* was left unpublished until more egalitarian historians took an interest during the nineteenth century, welcoming Díaz's claim that all the Spaniards who had been involved, great and small, deserved the credit. But this was still the story of a purely European victory.

When the Great European Powers were forced to surrender their American colonies, Europeans found themselves writing their own history as a story of loss and defeat. Writing about the conquest of Mexico, historians now characterized the Spaniards as foreign actors on an indigenous American stage. They read the Spanish accounts again and noticed that Cortés and his men were only part of the story. They had in fact managed to capture the Aztec capital at Tenochtitlán, later renamed Mexico City, only because they were supported by coalition of indigenous armies drawn from a population long subjugated and persecuted by the Aztecs.

As the great colonial empires broke up and slavery and segregation were abolished, American and European societies became home to the diverse many rather than the homogeneous few. Today, history can no longer simply be the origin myth of the Christian white man alone and must now explain the origins of our world to citizens from very varied backgrounds. But this situation presents a new set of problems.

Historians need facts with which to fill the blank pages of their histories, and although archaeologists and other students of the past can help, written documents have always been the historian's staple. We need the accounts of events written by the protagonists themselves, the official reports written by others at the time, the histories written by contemporary commentators and historians; we need the wealth of detail to be gleaned from legal documents and mundane account books. At the time of Cortés's conquest of Mexico, those documentary sources were almost all produced by Spaniards. How are we to write an accurate and balanced history of the events of that period when the Aztec Mexicans themselves left so few accounts of their own history? And it is more difficult still to write about Native American Indian history, which was an oral culture that did not produce written sources.

Similarly, the history of the Africans who served the Spanish Empire is not easy to write, because the sources were written by the masters and not the slaves. But, buried beneath the surface of the historical sources there lies a fragmentary, uncertain African-American history, and the subject of this book, Esteban, is one of the few examples of a sixteenth-century African slave whose achievements were so outstanding that it is possible to piece together his story from the contemporary Spanish documents.

Esteban became the pivotal character in the amazing adventures he and his Spanish companions lived through during the first crossing of North America in recorded history. The story is well documented because an official report based on the testimony of three Spanish survivors—Alvar Núñez Cabeza de Vaca, Andrés Dorantes de Carranza, and Alonso del Castillo Maldonado—was compiled in 1536.

How foreign and unfamiliar these names seem to us today, even though the first crossing of lands that later became the United States of America is so well documented. Nothing, perhaps, could be more indicative of the biases in our traditional understanding of when and where American history began. Long before the Pilgrim fathers established their colony, long before any significant European settlement on the east coast, Esteban had become one of the greatest explorers in the history of North America. In due course, he led a Spanish expedition deep into modern Arizona and New Mexico, and he may have died within the frontiers of the modern United States. Esteban was the first great explorer in America; he was also, perhaps, the first African-American.

In 1902, a distinguished African-American veteran of the Spanish-American War, Major Richard Robert Wright, concluded an academic paper by asking why Esteban had "remained practically in obscurity for more than three and a half centuries." "The answer is not difficult," he replied, for "until recently historians were not careful to note with any degree of accuracy and with due credit the useful and noble deeds of the Negro companions of the Spanish conquerors, because Negroes were slaves, the property of masters who were supposed to be entitled to the credit for whatever the latter accomplished. The object of this paper is to direct attention to this apparent injustice." He went on to remark that "if someone more competent will undertake a thorough investigation of the subject the purpose of the writer will have been accomplished."

I am considerably less competent than Major Wright to undertake the investigation he called for so long ago, but I am better placed to do so, for I have the benefit of a century of research by others, access to the Spanish archives, and time afforded me by a generous publisher. That my investigation has been as thorough as Major Wright would have wished is unlikely, for his were the exacting standards of his age; but more than 100 years later, it seems that this historical account of Esteban's life is still long overdue. In order to understand how Esteban came to be where he was, go where he went, and do what he did in America, we need first to learn something of how Spain came to have such a powerful empire in the New World.

In AD 711 an Islamic army crossed the narrow Strait of Gibraltar, a channel only eighteen miles wide separating Spain from Morocco—Europe from Africa—and quickly overran the weak Spanish kingdom. In the mountains of the far northwest, a few Christians resisted these invaders, but Islamic Spain soon developed into a great intellectual and artistic culture and its principalities and caliphates became centers of religious and political tolerance. Still, occasional periods of puritanical fanaticism led to social unrest and cultural censorship. During those periods of trouble and conflict, the tiny Christian kingdom in the far north was able to expand its territories and grow in strength.

The relationship between the Christian north and the Muslim south was complicated. Mostly, alliances were formed and broken with no regard for religion. But the call of the crusade or jihad was a potent political force, and from time to time border conflicts degenerated into outright holy war. More often than not, the Christian Spaniards won those wars, so that eight centuries after the Islamic invasion, almost all Spain was Christian once more. Finally, in 1492, Granada, the last remaining Islamic kingdom, capitulated to the Catholic Monarchs, as the joint King and Queen of Spain, Ferdinand and Isabella, were known.

The character of Christian Spain had been defined by this long history of *reconquista*, the "reconquest." It was a history of bloody battles, border raids, hostage-taking, and ransom. Spain was a land of warlords, overmighty aristocrats who won their wealth by violence and who ruled their estates with an iron fist. They were proud of their warrior status and their pure Christian heritage. Blond and blue-eyed, they despised work and commerce, the business of peasants, Jews, and Muslims. But there was no more of Spain to conquer. They had, quite literally, reached the sea.

In the crusading euphoria surrounding victory over Granada, the rulers of Spain decreed that the Jews should convert to Christianity or be expelled. Legend has held that money stolen from these refugees paid for Columbus's voyage. That tradition is largely myth, not fact, but Columbus was able to manipulate the triumphant spirit of the age in order to persuade the Catholic Monarchs to support his proposed voyage to China across the Atlantic. He had been searching for support for this scheme for ten years or more, but no one had ever taken him seriously because he had badly miscalculated the size of the world, as his detractors well knew. It is a curious fact, but on Columbus's own map, there was quite simply not enough space for America to exist. Where Florida is today, Columbus expected to find Japan. Suddenly and quite unexpectedly, the Spanish Crown gave its backing to a sailor most people believed to be a madman.

Because we have a copy of his journal, Columbus's first voyage to America is well documented. We know, for example, that when he landed on Cuba, on November 6, 1492, he sent his ambassadors deep into the interior of the island to contact the Chinese emperor. Nearby, he believed he would find India and the Spice Islands.

For decades, the Spanish discoveries in the Caribbean were largely disappointing, a series of islands populated by poor, primitive peoples and offering few signs of wealth. Although the great continental mass promised more, early explorations failed to find anything of real value. But Cortés's conquest of Mexico completely changed European attitudes to America. At the Aztec capital, Tenochtitlán, the conquistadors found a sophisticated civilization that was unimaginably rich in gold. The story spread fast of the adventurers who had been humble peasants and farmers in Spain and were now rich noblemen in an exotic land. But their position was precarious from the outset because Cortés and his men had disobeyed a strict order from the governor of Cuba that they should trade only on the mainland coast and that both settlement and conquest were prohibited. Cortés resorted to a legal trick to circumvent this problem, using a medieval law to turn his temporary army camp of flimsy tents into a town. This town was a bureaucratic fiction, but Cortés and his men could argue that the town council was legally independent of Cuba and could therefore authorize his campaigns in Mexico. Cortés and his men were well aware that while they had seized their golden prize from the Aztecs, their authority to do so was questionable under Spanish law.

In 1520 they were forced to defend their ill-gotten gains. That year, an experienced military captain and Caribbean slave trader, Pánfilo Narváez, was sent with a powerful army to arrest Cortés. But Narváez was to become one of the first men to experience the true power of Aztec gold. As the two Spanish armies prepared to do battle on Mexican soil, Cortés ordered some of his most trusted men to secretly cross the battle lines. He sent them armed with glittering gifts of unrivaled value and outlandish stories about the wealth of Tenochtitlán. Narváez's men were quickly seduced, his army deserted him, and Narváez himself was taken prisoner. Cortés kept his fellow general in a "gilded cage," regaling him with fine foods and luxurious living, but he took his prisoner by force to Tenochtitlán to see the splendor for himself. In due course, Narváez returned to Cuba defeated, but restless with greed.

On Cuba, the captains of slaving ships sent to raid for Indians on the northern coast of the Mexican Gulf reported improbable rumors of other wealthy cities deep in the heartland of the great swath of territory that today stretches from Florida to Arizona. Narváez hurriedly sailed for Spain, and in 1526 he obtained permission from Charles V to explore and settle the north coast of the Gulf of Mexico. He was given the title of Governor of Florida—Florida being the name the Spanish gave to the whole of the American south, from

Georgia and Florida to modern New Mexico. Overnight, he became the legal European ruler of an immense, but unexplored, world.

Narváez's expedition landed at Tampa Bay at Easter 1528, and in June he headed inland with 300 men. They ran into trouble almost immediately. Their progress was obstructed by mangrove swamps, they were harried by hostile Apalachee Indians, and they were ravaged by disease. Narváez's men were soon forced to slaughter their horses for food while they constructed five inadequate boats in a desperate attempt to reach Mexico by sea. When they finally set sail, the tiny, overloaded vessels were blown about and washed this way and that in the winter storms. Narváez himself was last seen shouting from his boat, being dragged out to sea, accompanied by a young page boy. Many men drowned; others were ruthlessly murdered by the coastal tribes of Louisiana and Texas; others fought among themselves and in their desperation descended into cannibalism.

Only four men survived this tragedy: Esteban, Cabeza de Vaca, Dorantes, and Castillo. Eight years later, in 1536, they told an astonished audience in Mexico the remarkable story of how they had survived by becoming shamans and bringing peace to the Indian tribes who revered them and how in return they were treated as gods. Their story came to epitomize for their contemporaries the struggle between good and evil at the heart of the Spanish Empire. They inspired Christian missionaries to preach a peaceful and spiritual conquest, but their reports also fired the fiercely greedy imaginations of the powerful conquistadors who competed for control of Spanish Mexico.

THIS BOOK TELLS the story of how Esteban came to be central to the survival of his companions and how, as a result, he was appointed by the viceroy in Mexico as the de facto military commander of the first Spanish expedition to explore deep into Arizona and New Mexico in search of the mythical lands of Cíbola and the Seven Cities of Gold.

Esteban's history is, in part, the story of how that first extraordinary crossing of America came to be told and written, first by the Spanish survivors of the expedition and then by later historians. It is also the story of how the history of that strange expedition to the Seven Cities of Gold came to be written.

In telling those stories, this book shows how Esteban was marginalized in the contemporary Spanish documents because he was a slave. As a result, he has been ignored or misunderstood by historians ever since. As well, in showing how it is possible to reconstruct Esteban's biography, this has become a book that asks questions about the ways in history may legitimately be written.

So this is a book about history and storytelling. Part One begins at the point when the four survivors of the Narváez expedition—Esteban, Cabeza de Vaca, Dorantes, and Castillo—began to tell the outside world the story of their first crossing of North America. It begins when they are reunited with Spaniards in northwest Mexico and traces their onward journey to the Spanish frontier outpost at Culiacán and then on to Guadalajara and Mexico City. Their knowledge of the previously unexplored north made them important players in the politics of Mexico City and the Spanish Empire, and that role determined how they developed and embellished their story. Part One ends as they complete their official testimony to the authorities in Mexico City.

Esteban's personal history begins in Part Two, which looks at the evidence for who he was and where he came from. It seems likely that he left North Africa during a period of terrible hunger, only to reach Spain while it too was suffering a devastating famine. We know that Esteban must have passed through the great Spanish port of Seville on his way to America, and Part Two ends with a description of that city during the period when Esteban was there.

Part Three returns to storytelling and examines the documentary evidence. First, it looks at the important difference between the two main versions of the story of the first crossing of North America which have survived and which are available to us: one written by a survivor, Alvar Núñez Cabeza de Vaca; the other by a contemporary historian, Gonzalo Fernández de Oviedo.

Part Four tells the story of Pánfilo Narváez's failed expedition to Florida, describing the preparations in Seville, the voyage across the Atlantic, the arrival in the Caribbean, and the landing on the Florida coast at Easter 1528. It describes the slow destruction of Narváez's army and Esteban, Dorantes, Castillo, and Cabeza de Vaca, tracing their travels along the Gulf Coast, through Texas, up the Rio Grande, and across New Mexico and Arizona. Part Four ends when they first make contact again with the Spanish world in 1536.

Part Five describes how, in 1538, in Mexico, Esteban was appointed as a guide and military commander of an expedition sent by the viceroy to search for the mythical Seven Cities of Gold, which were believed to lie somewhere north of the lands he had explored with Cabeza de Vaca, Dorantes, and Castillo, deep into Arizona and New Mexico.

Part One

▸◂

STORYTELLING
1536

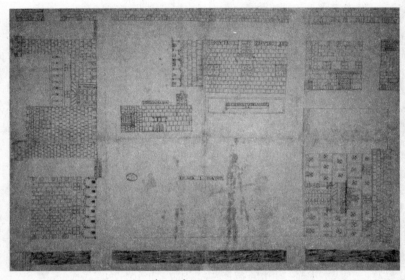

The Plaza Mayor of Mexico City (c. 1551). This diagram shows the imposing, well-defended buildings constructed at the heart of the city, the *traza*, where only Spaniards and Africans were permitted to live. (*Courtesy of Archivo General de Indias; MP México 3*)

I

JOURNEY'S END

The First Crossing of America, 1536

IN THE FAR northwest of Mexico, a posse of Spanish cavalrymen was riding deep in Indian country. History has marked these men as among the most bloodthirsty and brutal of the notoriously cruel conquistadors, the soldiers of fortune who forged the great Spanish Empire in the Americas. It was about the time of the spring equinox in 1536, the "ides of March," long believed to be a season when the Fates might conspire to destroy a man.

They had ridden far beyond the frontiers of their own world in search of peaceable Indians to capture and sell as slaves. A wave of fear had broken across the country. The inhabitants had fled high into the Sierra Madre or had taken refuge in the thick brush. No one had tilled the soil; there were no crops; the desert plain was barren; the fertile river valleys were emptied of people.

The conquistadors now reaped the harvest of those seeds sown in wrath and greed. Men and horses were weak with hunger. For days they had found no victims to enslave in that abandoned world, and they had no idea where they were. There was no one to guide them, nor lead them to water, nor give them food, and little grazing for their horses. They were hungry and thirsty, lost in the network of Indian trails that cut through the impenetrable backwoods of brush and thorn scrub, cactus, mesquite, and ebony, which closed in claustrophobically around them. The threat of ambush was unrelenting and terrifying in a land where the Indians used arrows poisoned with the sap of venomous trees. Their Mexican foot soldiers were restless.

This brutal band was led by a man called Diego de Alcaraz. He was hard and brave, a frontier man who lived in the saddle and a pirate who pillaged the land of its people. His creed was violence and his motive was greed. He cared nothing for the love of God, nor for the decrees laid down by his sovereign and the laws of Spain. He did his evil work far beyond any Spanish imperial jurisdiction, riding time and again deep into Indian country in search of peaceful victims he might easily enslave and send for sale in Mexico City.

But Alcaraz was worried that he had forced his posse of tough riders too far, farther north than any Spaniard had ever ridden before. With his men worn down by hunger and fear, he ordered a retreat, and they set out on the slow march south to the remote imperial outpost of Culiacán, the tiny military base they temporarily called home.

Culiacán was the farthest settlement on the most distant frontier of the vast Spanish Empire, a rough settlement on the green and pleasant banks of the San Lorenzo River. It marked the northern limit of a rich, fertile province known as New Galicia, ruled over by Nuño Beltrán de Guzmán, by almost all accounts a "natural gangster" and one of the most merciless and pitiless of men. Yet like so many conquistadors, he was as audacious, dynamic, and brave as he was bloody.

Alcaraz was cast in Guzmán's image; they were two psychopaths with a common, perfidious ambition. They sought out violence for violence's sake. They raped and pillaged, and with their plunder and the slaves they took, they could afford the cost of further violent missions. It was this cycle that had led Alcaraz and his men to overstretch themselves in the remote region they now hoped to find their way out of.

They backtracked for a week until they reached the lush and verdant banks of the Sinaloa River, where Alcaraz ordered his men to set their camp. This was some respite, at least, from the endless flat plain, covered in scrub. Some of the men thought they recognized the river. Early the following morning, he sent his most trusted man, Lázaro de Cebreros, to search for the trail to Culiacán, or for someone who could guide them there.

As Cebreros set out with three companions, he no doubt pondered the recent past. Only three years earlier, another slaving expedition to the region had found Indians wearing jewelry of horseshoe nails and other European objects. One man even had a scrap of material from a cape.

"Where did you get these things?" the Spaniards had asked. Eventually, they were told a grim story. A troubled Spanish ship had put into a nearby harbor, the crew had come ashore in search of succor and safety, and every one of them had been mercilessly massacred by Indian warriors.

Now, in the cool of early morning, as Cebreros and his companions went about their task, they suddenly tensed. Alert with fear, they sensed danger moving in the bush. Soon, Indians appeared. Cebreros watched as a group of fourteen or fifteen men approached along the trail. Instinctively, the Spaniards reached for their swords.

Then, as the Indians came closer, Cebreros noticed that these usually beardless men seemed to be led by a strikingly hirsute African, tanned deeply black by the relentless sun. Close behind him was a European, his blond hair and long beard bleached almost white. Both wore feather headdresses and carried the sacred rattles of Indian shamans or medicine men. Rude tunics sewn from deer pelts half-covered their nearly naked bodies. They went unshod, their feet deeply lined and cracked.

The two groups stared at each other a while. Then the blond man stepped forward. "Take me to your leader!" he ordered. He may have looked sinister, with his lion-like mane and Indian clothes, but he spoke with the familiar accent and arrogance of an Andalusian aristocrat.

Cebreros and his men were dumbfounded, struck silent by such a strange and improbable meeting.

These two men were Esteban, an African slave; and Alvar Núñez Cabeza de Vaca, a Spanish nobleman. One or two days' journey back along the trail were their two Spanish companions: Andrés Dorantes, Esteban's legal owner; and Alonso del Castillo Maldonado, a doctor's son. These four men were the only survivors of the disastrous expedition of 300 would-be conquistadors who had landed at Tampa Bay eight years before, filled with confidence that they could conquer Florida for Spain.

At this meeting with the four Spanish cavalrymen, Esteban and Cabeza de Vaca brought to a conclusion a remarkable odyssey. It is one of the most symbolic and momentous events in American history. They had completed the first crossing of North America made by non-Indians, the first crossing of North America in history.

These four strangers had walked from Texas. They had followed the Rio Grande to El Paso, rounded the Sierra Madre, and crossed through New Mexico and Arizona. From there, they descended into the Sonora valley of northwest Mexico, four godlike shamans who "came naked and barefoot" from the east. With them were a crowd of Native Americans, who flocked to these alien medicine men in search of hope in troubled times.

Cabeza de Vaca later recalled that as they had walked along the ancient trails of the southwest, they had "come across a great many diverse languages, through which God showed them his favor, for they always understood those languages and the Indians understood them. And thus, they conversed through signs so well that it was as if they spoke one another's languages." And, as they wandered among these myriad nations of many tongues, Esteban had "always spoken to the Indians, gathering all the other information they needed. He found out about the trails that they wanted to follow and what nations, tribes, and settlements might be thereabouts." Esteban was always the ambassador, a spy and a scout, the advance guard, the diplomat who dealt with the Indians while the Spaniards were mostly silent.

This, the Spaniards claimed, gave them great authority and gravitas, and as a result they commanded much respect. But although their mysterious silence may have created an illusion of supernatural power, it was clearly Esteban who controlled that power by communicating with the Indians.

Even as the Spaniards claimed for themselves the credit for their survival, their accounts betrayed the simple truth that Esteban had been their savior. Brave and resourceful, Esteban negotiated with their Indian hosts, always arranging the progress from one tribe to the next that led the four men across a continent and into Mexico. But Esteban needed his Spanish companions too: without them, he would have been a wandering alien; with them, he was a herald of the gods. All four men were locked together by a tight, triangular dynamic: Indian, African, and European.

Cebreros and his men remained silent for a long time. They looked at the strangers. They slowly realized who these peculiar apparitions were. They dismounted. Cebreros tenderly embraced Cabeza de Vaca, official treasurer of the long-lost Florida expedition. The little group of newly reunited Christians now made their way back to the riverside camp.

Esteban no doubt noticed the chained Indian prisoners destined for slavery. He was back among Spaniards, a slave once more.

As the twenty or so amazed Spanish cavalrymen examined the newcomers, Cabeza de Vaca and Esteban introduced themselves, still fearful of the slavers' possible response. But Alcaraz had little time to worry about how they came to be there. He was a man of action and he was in trouble. He confessed that for many days he had seen no sign of Indians, that he was lost, and that his men were going hungry.

Cabeza de Vaca was able to reassure Alcaraz that the necessary sustenance was at hand. He explained that ten leagues back along the trail Dorantes and Castillo were waiting for news. They had sent word into the surrounding lands for the Indian population to join them, and the Indians would have food.

This was welcome news to the desperate Alcaraz, who immediately ordered Esteban to lead three cavalrymen and fifty of his Mexican soldiers to search for Dorantes, Castillo, and their allies. They set out at once.

ON THE BEWILDERING trails, Esteban held the upper hand. At any time he might have disappeared into the bush to escape or organize an ambush. The Spanish riders and their Mexican foot soldiers were vulnerable, exposed to attack, and blinded by the dense vegetation. But instead, Esteban chose to lead this party of potential enemies into the heart of the four survivors' massed crowd of followers. Why?

Dorantes and Castillo soon heard from their scouts that Esteban was on his way with a posse of foreigners. At first, they must have been overjoyed. They must have begun to think of home and imagine what life would be like among Spaniards once more. But they would have quickly realized that they were far from safe. They had already seen the consequences of Aclaraz's brutality and knew well that such slavers could see them as rivals and might easily murder them. They had to gain the upper hand in any confrontation, and no doubt they now prepared a warlike welcome, calculated to intimidate the Spanish slavers.

Esteban, ever the advance guard and ambassador, must have approached the camp first, arguing in favor of caution. He quickly summarized the slavers' plight and pointed out the power the four survivors had in being able to control the Indians and therefore the food supply. Esteban, it seems, had now decided to throw in his lot with the three Spanish companions who had accompanied him from the Atlantic to the South Sea and once again submit to a Spanish world in which he had always been a slave.

Meanwhile, back at the slavers' camp, Cabeza de Vaca began to tell his story. As the last living royal official of the Narváez expedition, he requested that Alcaraz give him formal testimony of the year, month, and date of their meeting. He insisted that this be accompanied by a description of the manner of their arrival. Alcaraz complied with these bureaucratic requirements and then responded in kind, demanding that Cabeza de Vaca account for his bizarre attire and how he came to be there. In due course, he would interrogate all the survivors, insisting that they explain how and why so many Indians followed them in peace and harmony.

For all that Cabeza de Vaca's storytelling had already begun, the Spanish slavers can have only half-believed that half-recounted tale. How could he claim to have traveled through so many different lands where so many different languages were spoken by so many different peoples? They must have wondered what had really happened to him, and they must have doubted the strange black African would ever return.

But five days later, Esteban, Castillo, and Dorantes appeared at the head of the peaceful Indian army of 600 men and women that they had raised. There could be little doubt that something strange and yet quite wonderful had happened. The force of Indians now amassed before the little band of Spanish slavers with their Mexican servants and their Indian slaves. It may have been an army that seemed to come in peace, but it far outnumbered them and was also clearly a horde equally ready for war. These Indians had fled their homes in the face of Alcaraz's brutality. Their revenge could prove terrible. There was no place for faint hearts, for weakness might be fatal.

Alcaraz was audacious and brave, as well as brutal, and where others might see an army, he also saw potential slaves. He asked them to send word to the people who had lived in the settlements along the riverbanks, but were now in hiding in the bush, that they should bring food to the camp.

The four survivors did as he asked and sent their messengers to take word into the hinterland. Soon, a further 600 Indians arrived—men and women, children and infants. They brought all the corn they could manage, carrying it in cooking pots, which they had used to bury their staples to hide them from the slavers.

Then, as soon as Alcaraz and his men had eaten their fill and looked after their horses, a great argument broke out.

"We quarreled bitterly with the Spaniards because they wanted to enslave the Indians we brought with us," Cabeza de Vaca claimed. "We were so angered by this that when we left we forgot to take many of our bows and arrows with us and also some bags containing five emeralds, which we therefore lost forever. We gave the Spaniards plenty of hides and other things and we worked hard to persuade the Indians they could return home safely and sow the maize crop. But they wanted to go with us until they could leave us with another tribe of Indians, which is what they usually did. They were afraid that if they went home without doing so they would die and that if they went with us instead then they had no need to fear the Christians with their weapons."

This annoyed Alcaraz, who spoke to the Indians through his interpreters. "These four men are of the same nation and race as I am," he explained; "they are Christians and Spaniards, but they have been lost for a long time and are unfortunate people of little importance. I am the master of this land and you must obey me and serve me."

"But," Cabeza de Vaca went on, "the Indians gave little or no credence to all this. Instead, they began to talk amongst themselves, saying that the Christians must be liars, because we four survivors had appeared from where the sun rises, while the slavers came from where it sets. And, while we cured the sick, they murdered the healthy.

While we came naked and barefoot, they wore clothes, rode horses, and had weapons. While we coveted nothing, but instead gave away everything that they gave to us so that we were left with nothing, these others seemed to have no other purpose in life than to steal everything they could lay their hands on and never gave anything to anyone."

Cabeza de Vaca is without doubt responsible for embellishing this dialogue between Alcaraz and the Indians, polishing the rhetorical language for the sake of contrast and editing out much extraneous material. Even he admits that he had to rely on interpreters himself to understand what was going on. But the gist of that exchange is entirely plausible because it is ostensibly true.

The four survivors were almost certainly in a position to order their Indian army to confront Alcaraz's weakened force. Instead, they thanked the Indians for all that they had done and advised them to return to their former way of life, to go back to their homes and villages without fear. But according to Cabeza de Vaca, no sooner had they sent those Indians away in peace than Alcaraz ordered custody for the four survivors and sent them with Cebreros to Culiacán. There can be little doubt that the four men at least acquiesced or perhaps colluded in this decision, for their following of Indians was formidable indeed and had they chosen to engage Alcaraz, their army should surely have prevailed.

Alcaraz had judged the Spaniards and Esteban well. They may have been aghast at his treatment of the Indians, but they were more eager to begin the long march for Mexico. For all their fine words about protecting the Indians who had worshipped them as gods and revered them as shamans, the four now abandoned their loyal followers to the whim of some of history's hardest men.

"As we took leave of the Indians," Cabeza de Vaca wrote, "they told us they would do as we told them and they would return to their villages if the Christians let them. And so I say and affirm that it is quite certain that if they did not do so it will have been the fault of the Christians."

About 500 of these Opata and Lower Pima peoples who had followed the four survivors, and who were now far from their homes in the north, chose to settle nearby, below Sinaloa at a place called Bamoa. They took to wearing mother-of-pearl crucifixes tied in their hair, hanging down across their foreheads, in imitation of the four survivors, and they took to telling titillating tales of fabulous emerald mines in the lands of their ancestors. Many of them would return to those ancestral lands before long, with Esteban as their chieftain once more, leading them back north in search of the mythical Seven Cities of Gold.

With the four survivors out of the way, Alcaraz and his men seized many of these peaceful peoples as slaves. The four survivors were able to turn a blind eye because Cebreros and his three cavalrymen hurried his willing captives along the backwoods trails, so that they should not be witness to the slavers' crimes, nor have word from their Indians, or so Cabeza de Vaca later claimed. But Cebreros had no knowledge of the back trails to Culiacán, and for three days they traveled south, uncertain of their route, lost and without water. Seven of the Mexican foot soldiers lost their lives and many Indian slaves were abandoned in the bush before one of the four Florida survivors discovered water and saved their captors. That this hero remains unnamed in the contemporary sources is evidence enough that he was Esteban. Had it been one of the others, then he would surely have claimed the credit.

AFTER A FEW more days of hardship, Cebreros and the four survivors emerged from the high, desiccated brush into a wide, fertile plain of grasses and trees, where they easily took quail, hare, deer, and other game. The early harvest was coming to an end and the fields were filled with broad-shouldered, well-built Indians and handsome women, toiling in the searing sunshine. The roads were busy with people carrying loads of beans, corn, avocado pears, and squash back to their settlements on the banks of the river. From time to time they struck up a conversation, for by all accounts, these were a very friendly people. Young and old showed a happy curiosity toward the strangers, offering them sweet fruits, plums, guavas, and melons from the loads they carried on their backs. No doubt Cebreros and his men looked on, puzzled by the social ease of their countrymen among the "savages," whose eyes usually stared blankly out from scarred, unnatural eyelids and who wore mother-of-pearl and coral earrings, and had many colored feathers in their hair, all plaited with colored ribbons.

This peaceful world through which the four survivors now traveled bore witness to the ability of a man called Melchior Díaz, the commander of the garrison at Culiacán. He had established an uneasy truce in a few of the farming districts around the Indian town of Culiacán, but the four survivors soon discovered that, beneath the scenes of bucolic bliss, tension and discontent burned.

Cebreros left them at the Indian town and went on to the nearby Spanish settlement, where he reported to Díaz that although he and Alcaraz had failed to find gold or other precious minerals, they had found a prize that might perhaps prove more valuable still. They had, he explained, found four survivors of Pánfilo Narváez's long-lost expedition to Florida. He gave a garbled version of their story. But, he said, more important than anything else, they had an almost supernatural authority over the Indians, who flocked to them as sheep to a shepherd. With them around, the wolf need only wait at the door to the fold.

If we believe one rather unreliable source, Díaz must have been both overjoyed and astonished, for it seems that he had known the survivors back in Spain. That very night, he hastily organized a reception for these long-lost friends, a welcome as befitting as he could manage for men who had been exiles so long. Then he rode out to greet them personally, perhaps taking with him a member of his garrison from the Spanish city of Salamanca, a man called Maldonado who was probably some cousin of Castillo—whose mother's name was Maldonado and who was also from Salamanca.

As soon as Díaz arrived at the small Indian settlement where the four were waiting, he dismounted, and the old friends embraced and wept together, praising God for his mercy in sparing the only survivors of the expedition. "My house is your house," Díaz no doubt said, in the quaint, yet formal way still used by some old-fashioned Spaniards to this day. He offered them all that he could and welcomed them on behalf of the Governor of New Galicia, Nuño de Guzmán.

In that rich agricultural garden teeming with game, where the rivers were full of fish, the four survivors were presumably regaled with a great banquet, washed down with wine and mescal made from the maguey cactus. As the four relaxed for the first time in the friendly and familiar company of a Spanish military man, it is unlikely that Esteban felt much of an outsider. Successful captains like Díaz knew the value of an experienced man out in the field and Esteban was a quite different character from the slaves who had recently been flooding in from Africa, the *bozales*, men and women seized by the Portuguese on the African coast and carried straight to Mexico. It is unlikely that anyone, least of all Esteban himself, would have seen him as somehow similar to those new slaves who spoke no Spanish, knew nothing of Spanish ways, and knew nothing of America. Esteban was still an outsider among those Spanish aristocrats, but as yet there was little opportunity for that difference to be felt.

In the convivial atmosphere of this meeting, Díaz turned to his guests and, "with a great show of sentiment," he apologized for the behavior of Alcaraz. Had he been there, he insisted, the Indians would have been properly treated and the four would have been royally received. He no doubt began to explain that while his friends had been lost in the wilderness, Charles V had taken advice from the Pope about relations with the Indians. Concerned about his immortal soul, Charles had forbidden the enslaving of Indians. Díaz might be powerless to put a stop to Alcaraz at the moment, he explained, but the tide was turning; peaceful settlement was to be the new way. Guzmán's days, he may have whispered, are numbered. The four survivors were to be Díaz's ambassadors in Mexico City, carrying the message that Melchior Díaz was to be trusted as a man of peace among the Indians. Which is probably why he entertained the four survivors at this small Indian settlement, eight leagues from the Spanish garrison town where the illegal business of slaving continued as before.

The next morning, the four survivors perhaps rose late and engaged in easy conversation with the "affable" Indians while they enjoyed a breakfast of sun-dried squash in the shade of the woven palm-leaf porches. But as they were setting out to continue their journey to Mexico City, Díaz came to them with a proposition.

Stay a while, he pleaded, and you will be doing God and His Majesty a great service. Much of the countryside nearby is deserted, the fields are untended, and everything is going to ruin. The Indians have fled into the bush and refuse to return to their villages. You must send word to them, ordering them, in the name of God and the king, to come back to settle once again on the plain and return to their fields.

The four men were not really in a position to refuse, but they all knew success was going to be difficult to achieve. Their normal method of proceeding having been disrupted by Alcaraz and Cebreros, they had no Indian followers to announce their arrival.

Esteban was the great communicator of the four, the one who had always dealt directly with the Indians. He was used to marshaling the Indian messengers, men among his personal entourage who understood the culture and language of their neighbors. These native diplomats had always been the first to announce the arrival of the strange shamans as gods who had come from the east to heal the sick. Those Indian messengers always built the myth, recounting rumors of miracle cures and the raising of dead men. With that supernatural aura in place, Esteban could take the stage, decorated with the sacred objects of the spirit world, dressed as a medicine man, and begin to work his own kind of magic, the magic of his words, of the tale he told, the magic of narrative.

But without their entourage of Indians they had a problem, and it fell to Esteban to find a solution. He settled on two Indian captives brought down to Culiacán by Cebreros, men from the Sinaloa River valley who had seen the strange array of refugee Indians gathered around the Florida four. Chained and despondent amid the hunger and brutality of Alcaraz's encampment, these Sinaloans had sensed salvation among the massed men, women, and children who had arrived with the four survivors. In contrast to the constant tragic exodus they had known for years, they had now witnessed a reverential return to a land of Christian promises. There was hope in the wonderful stories of medical miracles, of these outsiders, these powerful, charismatic men who brought people together in the face of apocalyptic adversity.

Esteban sent these men as heralds to summon the fearful population. One messenger struggled high into the impenetrable mountain fastnesses where rebel leaders had regrouped. The other went back to the Sinaloa River and sought out the displaced people of that region. They carried the usual simple message. The foreign gods wished to meet with them near Culiacán. As a sign of his authority and sincerity, Esteban gave each of these messengers the most potent symbol of shamanism and political power available to him, a finely decorated rattle, a dried gourd with seashell rattlesnake tails, plumed with magical feathers.

For seven days, the four survivors waited in the Indian town as Melchior Díaz, their host, held them hostage to the fortunes of regional peace within his personal frontier fiefdom. He continued to keep them away from the Spanish garrison, perhaps to preserve their power and influence, but it was also a prudent measure, for he no doubt feared dissent and conflict if they came into contact with his motley collection of colonists.

Finally, the messengers returned. Despite the continuing brutality of the Spanish slave raids, three great Indians chiefs from among the army of resistance gathering in the mountains had agreed to visit Esteban and the Spaniards. These noble chiefs arrived in princely splendor, carried in hammocks by servants and accompanied by their noble retainers, fifteen men of rank and standing. And they came with gifts, sumptuous offerings of peace: brightly colored beads, turquoise, and feathers.

Melchor Díaz now took control, ordering the interpreter to speak to these chiefs on behalf of the four survivors. Tell them, he said, that the four were messengers from God in heaven and that they had walked the world for many years preaching to all the different nations and tribes, teaching them a belief in God and how to serve Him.

Díaz then ordered the interpreter to read out the "Requirement," the *requerimiento*. This peculiar document had its origins in some of the strictest interpretations of Islamic Shari'a law, medieval interpretations that had once encouraged Muslims engaged in holy war (jihad) to enslave non-Muslims who refused to accept Islam. Reinvented by Spanish bureaucrats for use in the New World, the Requirement must be one of the most ludicrous legal institutions ever conceived of by a lawyer or legislator. It was an official statement of crown policy toward subjects in the New World. All Spanish conquistadors were obliged to have the document read out loud in the presence of a notary and other crown officials whenever they formally took possession of new territory on behalf of the Spanish sovereign.

As a result, from the date of its institution, the Requirement was regularly read to empty beaches, forests, mountainsides, and deserted villages. Often it was read to utterly uncomprehending Indians who understood no Spanish and so had no idea of what it meant.

This was one of the relatively rare occasions when the Requirement was read using an interpreter. This at least made it clear that obedience was required and any defiance of the Spaniards would lead to persecution—a simple enough message. It also gave any Indians who cared for such things the chance to grapple with the improbable new theology of Christianity. That, it seems, is what happened next during the peace negotiations at Culiacán.

Esteban and the other survivors asked the Indians about what gods they worshipped and made their sacrifices to and which gods sent rain to their cornfields and which gave them health. They replied that they believed in a man who lived in the sky, called Aguar, who had created the whole world and everything in it.

How do you know of Aguar? the survivors asked.

The Indians replied that their fathers and grandfathers had told them. They had known about Aguar for many generations and they knew that he was the source of all the good things in life, such as water. Given that the Spanish word for water is *agua* and this discussion no doubt involved much pointing at the sky, it seems likely that the Spaniards and the Indians were heartily misunderstanding one another. This did not prevent Cabeza de Vaca from explaining that the four survivors then told the Indians that they should refer to Aguar as God and that they should serve him and adore him as the Spaniards instructed because to do so would be to their considerable benefit.

We can only guess at what the Indians really thought about the bizarre ritual of the Requirement and the peculiar theological discussion that followed, but they sensibly replied that they would be good Christians and serve God as instructed.

Esteban and his companions then ordered the Indians to come down out of their mountain hideouts and return to their lands in peace, harmony, and safety. They should rebuild their villages and set up a house of God among the new houses. In front of this church they should raise a cross like the one the Christians had in front of their church. Whenever Christians came to their settlements, they should greet them with crosses in their hands as a sign of peace. They should invite the Christians into their homes and share their food with them. In that way, it was explained, the Christians would do them no harm but instead treat them as friends.

Again, the Indians assented, agreeing to do all that the survivors asked of them. Melchior Díaz then treated these Indians well, giving them a number of blankets.

The delegation of Indian dignitaries then returned to the mountains to hold council about the meaning of these messages of peace. They held night dances or performed other rituals to make contact with their ancestors and gods, seeking out the wisdom of another world. Whatever they said and whatever form that debate took, the familiar crowds of friendly Indians began to stream into the small settlement the four survivors temporarily called home, bringing their accustomed offerings of shells and feathers, distraught pilgrims seeking help from the four strange aliens.

These Indian pilgrims must have celebrated their return to normal life, holding the traditional ceremonies at which they drank mescal and pulque, distilled or fermented from the agave cactus. At these ritual gatherings, all the people of a settlement gathered and those who had learned how to drink with the necessary gravitas would drink, while the inexperienced and the young watched.

Among this crowd of Indians were no doubt the medicine priests, shamans who communicated with the supernatural sphere, men who were the traditional tribal contact with the powerful spirit world, witch doctors and the keepers of tradition. They knew the secret healing properties of herbs and plants; they were skilled at sucking wounds and the magical use of fetishes to remove evil from the pain-racked bodies of the sick, skills in which the four Florida survivors had become well versed. The shamans surely came to meet the four foreign witch doctors who traveled the world in peace. To them it seemed a meeting of like with like, and some further ceremony must have taken place, some pagan ritual left off the record for the sake of Catholic orthodoxy. One later missionary described the night dances in which the chiefs dressed in terrifying masks, some as scorpions, others as snakes. Most fearsome of all were those wearing grotesquely carved pumpkins that covered the whole head.

Three years later, when Esteban returned, he formed such a strong and close friendship with these Indians that it aroused the jealousy of two monks who traveled with him. This is perhaps evidence that Esteban had long played his role in the night dances and the drinking festivals, perhaps taking center stage while his three Spanish companions remained aloof, their reticence a crucial element in the potent political dynamic of their different roles.

Whatever the true nature of these amicable exchanges may have been, in the written documentary record the four survivors simply reported that they had instructed these Indian pilgrims in the importance of constructing churches and the prominent display of the cross. They reported that they had told the chiefs and wise men to bring their wives and children to be baptized.

Melchior Díaz then swore in the name of God that he would neither organize nor consent to any armed campaign into the lands of the people who had been pacified; nor would any slaves be taken from among the people who had been assured of security and safety. He vowed to keep that pledge to God until he was ordered to do otherwise by his superiors.

Only then did Melchior Díaz and the four survivors set out for the Spanish township of San Miguel de Culiacán. There, Indian messengers delivered reports of many people coming down out of the mountains and returning to their villages and fields on the plains. They were building churches and raising crosses as instructed. For two weeks the four men waited at Culiacán, surrounded by the Spanish colonists, as news of their unexpected evangelical success arrived daily. The news was soon confirmed by an astonished Alcaraz, who arrived and reported to Díaz that the Indians had returned to their lands, bringing gifts of food and other things. These Indians, Alcaraz explained, had spent the night at his camp, and they had insisted so forcefully on being accorded the rights Díaz had granted them that it had unsettled him.

2

CAMINO REAL

The Royal Road to Mexico City, 1536

BY MAY 15, 1536, a caravan had been organized for the difficult journey to Compostela, the capital of New Galicia, seat of Nuño de Guzmán's government. Esteban, Cabeza de Vaca, Castillo, and Dorantes left Culiacán in the company of twenty cavalrymen and 500 Indian slaves. Despite all the documentary reports of peaceful settlement and Spanish promises to desist in the abuse of Indians, the stark reality was clear: slaving was still as much the business of Culiacán when the fours survivors left as it had been when they arrived. The twenty cavalrymen were to accompany the caravan forty leagues along the road until they were beyond the reach of the Indians from around Sinaloa and Culiacán who might rescue the captive Indians. Then, the twenty cavalrymen would turn back, leaving six Christian slavers and the four survivors to continue the further sixty leagues to Compostela.

At Compostela, a minuscule settlement with no proper church, no proper buildings, and none of the necessary fabric of civic government, Governor Nuño Beltrán de Guzmán received the four survivors with as much pomp as he could and pampered them for a dozen days. He arranged lavish hospitality for the three Spaniards in the barrack house of Captain Francisco Flores. He also offered them rich clothes from his own wardrobe, but the heavy Spanish attire was tedious and uncomfortable for men accustomed to the freedoms of nakedness and cotton blankets, and they turned down his offer. The four survivors were also no doubt keen to keep wearing their Indian garments, potent, startling symbols of their suffering and mysterious survival, trophies of their great journey with which to impress their fellow Spaniards.

Esteban was lodged in Guzmán's own paltry palace, where he was feted as rarely a slave has ever been regaled. There were gifts of food and perhaps the company of women from the harem Guzmán is said to have kept. We can assume that there was cajoling as well as carousing, as Guzmán pressed him for information about the unexplored north. There were threats too, for Guzmán was a governor general on the brink of disgrace and desperate to escape his enemies by leading a successful campaign of conquest in the north. But Esteban remained calmly reticent, unperturbed by his host's anger. From what Díaz had told the survivors at Culiacán, he knew that Guzmán's downfall was imminent. In fact, in due course, testimony provided by the four survivors themselves would assist in his prosecution. But for the time being, the four hurried on for Mexico City.

NEWS THAT FOUR men had survived Pánfilo Nárvaez's Florida expedition had already traveled fast. Settlers throughout New Galicia had heard something of the survivors' reappearance in the badlands of the north. Garbled versions of the story were carried by word of mouth, told and retold by Spanish, black, and Indian messengers alike, verbal traffic on the main highway to Mexico City, the *Camino Real*. Hungry and dissatisfied conquistadors heard the whispered rumors and speculation about the wizardry or even wickedness of that survival. It was a news-rich cocktail that fascinated them and stimulated their appetite to find out what had really happened.

All along the road from Compostela to Guadalajara and then on to

Mexico, settlers and colonists came to see them, to offer them hospitality, to hear their story at first hand, and to give thanks to God for their salvation. Then, somewhere near Guadalajara, it seems likely that they would have met three inquisitive settlers who were especially interested in their story.

These three men—Antonio de Aguayo, Juan de Castañeda, and Alonso de Castañeda—knew that Esteban, Dorantes, Castillo, and Cabeza de Vaca would likewise be particularly interested in the story that they had to tell, because they had also been part of Pánfilo Nárvaez's expedition. They too had all sailed from Cuba in quest of the fabled promise of Florida. They too had landed at Tampa Bay. But when the main body of the expedition marched inland in search of conquest, Aguayo and the Castañedas were among the 100 men and women who remained with the boats. The four survivors, Narváez, and the hundreds of men who perished had never seen those ships again, and that had sealed their fate.

Aguayo and the Castañedas explained that they had searched the endless expanses of coastal Florida throughout the summer and fall of 1528. Time and again they had clashed with hostile Indians during their sorties inland in search of the expedition. One of their number, an aging man called Juan Durán, later reported that during one desperate skirmish he was "gravely injured," as was his horse. Eventually, they realized that their search was futile. So, when the threat of winter storms had passed, they made preparations to return to Cuba.

Aguayo and the Castañedas were among a group of a few broken survivors who guided their ship into the harbor at Havana in the spring of 1529, a year after the main expedition had disembarked at Tampa Bay. Battered by hunger and the elements and harried by Indian braves, they sold a few Native American slaves on Cuba for what little the local merchants were prepared to pay. They took their money and went off elsewhere, to Mexico, to try their luck again in another as yet uncertain but surely less inhospitable place.

But there is some suggestion that the search for the expedition may not have been especially assiduous. Cabeza de Vaca reported that years later various witnesses told the four survivors that as soon as the main expedition set out overland, the women who were left behind all forsook their husbands and "married or began sleeping with the men who remained on the boats." A document I came across in Spain hints that this may have been true. It mentions that Mari Hernández, a woman who went on the Florida expedition with her husband, Francisco Quevedo, had married another man "who was also a conquistador" by 1539, but it does not say exactly when.

Moreover, the Indian captives brought back to Cuba in 1529 reported to officials that Narváez and his men were still alive, somewhere on the Florida coast, although their lives were basic. "All they do is eat, sleep, and drink," the Indians said. There is no record of the excuses offered by Aguayo, the Castañedas, and their companions for failing to find the stranded expedition in spite of the help they should have had from these Indian captives, but we do know that this news was enough encouragement for Narváez's loyal and determined wife, María de Valenzuela, to organize a further, tragically unsuccessful search. But Aguayo, the Castañedas, Mari Hernández, and these other unfaithful men and women took the money they had made from selling their Indian captives as slaves and sailed their damaged ship to Mexico. They arrived as "defeated soldiers," "naked and poverty-stricken." They were rescued by friends and acquaintances, men they knew from their old lives in Spain. Aguayo could count one brother among the dead during the conquest of Mexico and another brother among those who lost their lives during the futile exploration of Honduras. Of this luckless brood at least Antonio remained alive. He now turned for help to another immigrant from Portillo, the town in Spain where he was born, and he was offered shelter and what little succor was available.

These survivors had arrived in Mexico during a period of extreme political tension. In 1529, Nuño de Guzmán was still the governor, but he had made a mortal enemy of the powerful Archbishop of Mexico, Juan de Zumárraga, and he was preparing to flee Mexico City by leading an army into the unconquered lands to the north. For bankrupt men like Aguayo and the Castañedas, there were few opportunities in Mexico City, and they must have felt that their best option was to throw in their lot with Guzmán, whose villainy and brigandage asked little of a man other than a strong arm. They had ended up poor and desperate in Guzmán's ailing, badly run settlement at Guadalajara.

Aguayo and the Castañedas were no doubt determined to find out from the four survivors whom they had abandoned on the Florida coast whether there might be prospects of rich conquests in the remote north. But they also had more personal questions about the fate of relatives, friends, and rivals. No doubt there were some other men who wanted to be sure that their wives' former husbands were really dead.

Esteban, Dorantes, Castillo, and Cabeza de Vaca were all quickly learning that the story they had to tell might prove much more powerful than the truth about what had really happened to them. Everyone wanted news of the unknown lands to the north; the question was how best to manipulate their account of what had happened so as to turn that news into kudos, power, and wealth.

3

MEXICO CITY

July 1536

THE FOUR SURVIVORS were now living out the final days of an odyssey that must have often seemed an endless, directionless quest for home. They were now slowly becoming reaccustomed to the once familiar Spanish ways of the people around them as they neared the capital. When they reached the crest of a high pass and began their final descent, they marveled at the Valley of Mexico, spread out before them with its towns and cities that seemed to float on the wine-dark waters of the great lake.

Hernán Cortés and his men had been the first Europeans to see this valley, in 1519, and even that troubled band of hardened frontiersmen were awestruck when they reached this seemingly enchanted world. They asked one another whether the panorama before them might not be some scene from *Amadis of Gaul,* one of the earliest best sellers in the history of publishing, a late medieval romance of chivalry full of tales of derring-do, a rich recipe of fantasy, sex, and violence set in a make-believe world of knights-errant, princess-brides, and ogres.

In the long letters that Cortés sent to Spain, he described newly subject Mexico in such exuberant language that it unsettled the most powerful sovereign the world has ever known: Charles V, Holy Roman

Emperor; King of all Spain, of Castile, of Aragon, and of Granada; ruler of half Italy; heir to the Burgundian tradition of courtly chivalry; the Prince of Christendom who, in his gastro-diplomacy of public overeating and prodigious drinking, rivaled Henry VIII for gluttony and greed. But this fearless warrior, Charles, was perturbed by the personal triumph of this upstart Cortés, a dropout law student who had become, suddenly, the most powerful man in America. All Spain thrilled to the story of Cortés's conquest of Mexico and the powerful Aztec Empire with its unimaginable stores of gold and precious jewels. All Spain could read this story as a parable of revolution in which an intrepid individual defied authority and won a magnificent prize. It is perhaps a reasonable albeit excessive exaggeration to suggest that Charles needed Cortés's loyalty as much as Cortés needed the approval of his sovereign.

Cortés's letters to the emperor were published in Seville so all Europe might know that the great civilization found in the New World had capitulated to a handful of God's own Spaniards. In those letters Cortés wrote that the marketplaces of Mexico were vast, beyond comparison with even the massive market square of Salamanca in Spain. Like the bazaars of modern Marrakech or Istanbul, this gargantuan commerce was strictly ordered, street by street, each street being designated for the shops of jewelers, cloth merchants, metalworkers, basket weavers, potters, grocers, butchers, fishmongers, and the rest. In this great *tanguis*, as the Aztecs called this extraordinary exchange, where cocoa beans served as currency, tens of thousands traded with each other every day.

The panoply of produce found in this extraordinary plaza was also a feature of subsequent Spanish rule. One amazed visitor wrote that, all year round, the markets overflowed with fruits quite different from those of Spain: bananas, guavas, avocados, pineapples, an apple-like fruit known as *zapotes*, and prickly pears, to name but a few. There were fresh fruits familiar in Spain as well—oranges, lemons, limes—and in some places there were giant citrons and grapefruit. There were early figs and grapes, peaches, apricots, pears, and apples. And all seemed of extraordinarily good quality and very cheap. Esteban, Castillo, Dorantes, and Cabeza de Vaca, so used to hunger and hardship as they were, must have felt they were trespassing in Paradise itself.

The textile market of Mexico, Cortés claimed, was even richer and more sumptuous than the wonders of the silk market in Granada. The mention of Granada was doubly charged with significance for the Spanish monarch, Charles V. Granada was the Islamic jewel in his many crowns, a former Muslim stronghold that was for centuries an Oriental thorn in the side of Catholic Spain. It had long been a reminder of the 800 years that Spain was culturally part of the African, eastern, Asian, Islamic world.

But in 1492 the city had been recaptured by Charles's grandparents, the Catholic Monarchs. Suddenly, the cloud was lifted and the Spaniards lost their sense that a mysterious, alien world was a threatening presence close at hand. With the fall of Granada, Spain felt that the real, truly mysterious Orient might one day be hers to conquer.

Columbus had believed he was landing in China when he first set foot on the shores of Cuba in 1492 and had sent ambassadors inland to communicate with the Great Khan, but it was quite clear to Cortés that Mexico was not the east. Mexico was something else, altogether more mysterious. It was a place unheard of, an unsuspected, utterly unknown empire until Cortés's dazzled followers arrived. Europeans still believed, however, that America was part of Asia, and any news of a new and unimagined empire seemed like something out of Marcos Polo's fantastical stories of the Orient.

Montezuma, the Mexican Emperor deposed by Cortés, even had a rural pleasure complex with a zoo of jaguars, mountain lions, and brightly plumed birds, just as Marco Polo's Great Khan had his

Xanadu. Montezuma also kept a collection of assorted human oddities: dwarfs and hunchbacks, jugglers and acrobats. It was a magically surreal world for the Spaniards, a land of abundant gold and silver, precious metals prized for their aesthetic value alone. The Spaniards' blind and all-encompassing greed for such things was always a mystery to the Mexicans. This was a strange place indeed, almost a place of make-believe.

For those Europeans, like Charles V, who could only read reports of Mexico and listen to tales told by travelers from across the Ocean Sea, it was a land imbued with ideas of the Orient, ideas drawn from old traditions of a mysterious east. The indigenous people were called Indians, and America itself was known as to the Spaniards *las Indias*. For centuries, the encroaching military might of the Turk had cut off Europe from the magical world of China and India, and Europeans were consequently ignorant of the east. The legends grew more fanciful and more enticing as they filled that void of knowledge.

Esteban, Castillo, Dorantes, and Cabeza de Vaca now descended into the Valley of Mexico, a place they already knew about from Cortés's widely publicized letters. And, as they approached this promised paradise, they knew that their own reports of their adventures would have to be filled with similar wonders and marvels if they were to excite the unrealistic expectations of the Spanish officials who could grant them licenses to return to the north as governors and officers of the crown.

THE VALLEY OF Mexico is a wide volcanic bowl dominated by mountain walls and two fiery mountains, towering above all else, often smoking gently on the horizon. In those days, the city was lit by searing, almost equatorial sunshine, fierce in the thin clear air of altitude. That ethereal atmosphere perhaps seemed heavenly to the four survivors who had struggled up from the humid tropical coasts into the thin oxygen at 7,500 feet. The survivors could clearly see the towns and villages, seemingly floating on the waters of the lake, like enchanted island citadels, smoke rising from the chimneys of a thousand hearths. When Cortés and his Spaniards first crossed the threshold of the Aztec capital and gazed upon this extraordinary sight, they were astonished by the vast buildings of the residential suburbs that crowded the lakeshores.

Years later, the old soldier Bernal Díaz del Castillo remembered his sense of wonder and awe as he and his companions strode boldly along the dead straight causeway leading into the Aztec capital of Tenochtitlán itself. The Spaniards were all but lost for words with which to describe so strange a place, with its temple-pyramids and palaces, its defensive towers and turrets, all rising confidently out of the waters of the lake. And that is when they turned to *Amadis of Gaul* and their fantastical novels of chivalry to describe the otherwise indescribable reality of Mexico.

Mexico seemed a land of fantasy even to the rough-tough army of 1,500 Spanish pragmatists who made it their own in the face of impossible odds. *Amadis of Gaul* was the direct inheritor of the Arthurian tradition, with its stories of emboldened knights winning their spurs in a fantastical landscape populated by equally fantastical adversaries. It was a land of the imagination. Those Spanish pioneers in America felt that fiction had become reality. Mexico reminded them of the Castle of Bradoid, described in *Amadis* as "the most handsome castle in all the land."

Following the traditions of the chivalric genre, the author of *Amadis* had set the action in formidable surroundings. Bradoid was built on a rocky outcrop between a salt marsh and the sea. Like Mexico City, it could be approached only by boat, or by a long, broad causeway, wide enough for two ox wagons to pass. Where the water was deepest, this causeway was defended by a drawbridge, flanked by two tall oak trees. As the heroes of the novel approached, they saw two young maidens and a squire waiting beneath these trees.

It is little wonder that the utterly foreign world of the Aztecs, their empire, and their capital city, Tenochtitlán, should have inspired a literary reaction in the first Europeans to see it. We have no record of how Esteban, Cabeza de Vaca, Dorantes, and Castillo first felt when they contemplated Mexico City as they reached the end of their almost interminable peregrination through the swamps of Florida, the seas of Alabama, the waters of the Mississippi, the lagoons and prairies of Texas, the deserts and sierras of Coahuila, and the wild fertility of Sonora and Sinaloa. For the three Spaniards, Mexico no doubt offered a welcome vision of verdant fertility, a bucolic paradise that held little fear of hunger and starvation. There was activity in the fields around the shores and in the *champas*, the "floating" market gardens of anchored reed beds overgrown with trees and built up with mud and silt. These artificial fields were the breadbasket of Mexico, rectangles of fertile ground crisscrossed by waterways. These regimented islands were a cornucopia of agricultural greenery and Aztec flower gardens. Between the *champas*, the deep canals played host to a traffic jam of brightly painted canoes and barges.

But the Florida survivors' sense of being miraculously invited into a worldly paradise must have been tempered by concerns about the challenges ahead.

As Esteban entered this Eden, he must have wondered whether his return to the Spanish world and the subjugation of slavery might not be his own perdition. We can only guess at the trepidation he felt as he looked back on those happy days in hell which he had so recently shared with his Spanish companions among the Indians of the American southwest. Since then, he had seen the wild brutality of the Spanish slavers under the savage captaincies of Alcaraz and Cebreros and the more ordered civility of Melchior Díaz and Culiacán. He had met Nuño de Guzmán in Guzmán's private fiefdom at Guadalajara and had learned that his own experience among the Indians had made him a valuable man. Esteban knew only too well that he was a black

African and that Spaniards saw first the color of his skin. The European world to which he was now returning defined him as a slave, and his immediate future was clearly going to be very different from life among the Native Americans to the north. What, he must have wondered, would Mexico hold in store for him?

In the cold air of the mountain morning a mist rose from the Mexican lake and sunbeams played in the clouds of smoke rising from the fires and chimneys. The four survivors now entered the water-bound metropolis along the broad causeway where so many Spaniards, Africans, and Indians had lost their lives during the bloody battles of conquest. The lake had once run red with the blood of the fallen. But now the causeway led through the peaceful Indian suburbs where Mexicans went about their usual business or constructed parish churches under the watch of priests and friars.

The four survivors must have watched the scenes of seemingly harmonious industry around them with considerable wonder. Mexico was the largest city that any of them had ever seen. They noticed the black foremen, slaves themselves, who drove or directed gangs of Indian workers. They saw hummingbirds that darted hither and thither about the flower gardens. There were canoes loaded with abundant food.

The causeway led past a tiny chapel that marked the scene of especially bloody fighting during the conquest. Here they and their companions no doubt stopped to offer prayers for the souls of the dead and thanks for their own deliverance. There was no damsel to greet them, for this was no novel of chivalry. Instead, close by the chapel was an allotment belonging to Juan Garrido, a man who must surely have taken an interest in the four survivors, and certainly an especial interest in Esteban. Garrido was also an African conquistador and adventurer; he had been among the first to settle Puerto Rico in 1506 and the first to sight Florida in 1513, and he had participated in the conquest of Mexico in 1519.

Garrido was one of a few Africans who lived in Spanish Mexico with some sense of freedom, men who were technically and officially free. For all that they were trapped in an imperial cage, it was a social prison that they shared with many humble Spaniards, poor peasants and Jewish and Muslim refugees from the Inquisition in Spain. Such men and women, black and white alike, are all but lost in time, individuals who simply subsisted as unfortunates among the wealthy in a land of opportunity.

Among such flotsam and jetsam, washed hither and thither on the imperial sea of fortune, there were a few free Africans who fared well, if unspectacularly, in the rough, tumultuous waters of the Spanish colonial world. They are difficult to find in history books because black Africans are mentioned so infrequently in the original historical documents. They rarely petitioned the Crown for favors; they were almost never called as witnesses to support the petitions of others. They had no money, so no one sued them. They were powerless to take action against those who abused them. They left no written records.

Even when Africans do appear in historical accounts, they are, like Esteban, often misunderstood or misrepresented. Among such men and women, Juan Garrido stands out as a perfect example. Even though by his own account he had sailed to America as a free man, Garrido is almost always described as Cortés's "slave": simply because he was a *negro*, with few exceptions historians have assumed that he must have been a slave. And, because he fought in the conquest of Mexico, they have assumed he belonged to Cortés. He had in fact sailed for America as a free man.

When the four survivors arrived in Mexico, in 1536, Garrido was a relatively unimportant member of the tight-knit Spanish community, a veteran explorer and conquistador. For Esteban, Garrido was an example of what his own life might be like in Mexico should he be fortunate enough to be granted his freedom by Andrés Dorantes. And just as Esteban would have taken an interest in this retired adventurer, so Garrido would have taken an interest in Esteban.

We in fact know a lot about Juan Garrido, largely because he was one of the few Africans to petition the Spanish Crown. He began his approach to the authorities in 1538 in exactly the same way as many conquistadors who had fallen on hard times. He hoped that he would be offered some kind of pension or some other material benefit to make old age more tolerable. To do this, he first went to the Audiencia of Mexico, a cabinet of judges or magistrates who acted both as the executive and as a supreme court. In Mexico, the viceroy and the Audiencia together were the virtual proxies of the king and his court in Spain. The Audiencia was the seat of Spanish colonial government and the high court of Mexico. There, witnesses were called to testify on Garrido's behalf. A written record was made and a copy was sent to Spain for the King and his ministers to consider.

This document states that "in the great City of Mexico in New Spain on the twenty-seventh day of the month of September in the year of Our Lord 1538, Juan Garrido, black in color, a citizen of the aforesaid city appeared before Your Worship." It then records what he said: "I state that I need to demonstrate proof of how I have served Your Majesty in the conquest and control of New Spain since I came here with the Marquis of the Valley [Cortés], serving in his army. I was present during all the expeditions, conquests, and occupations that the Marquis ever prosecuted. All this I have done at my own expense without receiving any salary, nor lordship over any Indians, nor anything else, despite being married, as I am, and [despite] always having been a citizen of this city and living here.

"Moreover," he added, "I went to explore the islands in the South Sea [Pacific Ocean] where I suffered great hunger and hardship," by which he means he had been on the first expedition to California. He then goes on to say: "I was the first man in New Spain to make the effort to sow wheat and to see if it would grow here, which experiment I did at my own expense," and he adds that he had also been the first to plant "other things" as well, a claim supported by Cortés's biographer Gómara.

A document I uncovered at the Archivo de Indias, in Seville, shows that once Garrido had completed these formalities, he then traveled to Spain in order to present his petition at court in person. It seems that to offset his travel costs, he had somehow persuaded a free Mexican Indian called Pedro to go with him. As soon as they reached the Spanish court, the unscrupulous Garrido made numerous attempts to sell the unfortunate Pedro as a slave, but he was prevented from doing so by a number of Mexican citizens who testified that Pedro was a free man. Once Garrido had finished his business at court, he set out for Seville in order to take ship for Mexico again. At Ciudad Real, an important city on the road to Seville, he finally managed to sell the unfortunate Pedro to an unsuspecting slave dealer. In time, Pedro successfully protested to the courts, which is why we have a record of these events—but by then Garrido had long since escaped to Mexico, and it seems justice never caught up with him.

We can only imagine that the four survivors stopped to speak briefly with this ingenious if picaresque African gardener as he tended his allotment by the chapel on the causeway. An inquisitive man, Garrido no doubt put down his hoe or his trowel and introduced himself to these men whose exploits had already made them famous in the city. Wherever and whenever he and Esteban met, we can be as good as certain that they did meet, for it can hardly have been otherwise in the close-knit world of the black conquistadors of Mexico City.

THE CAUSEWAY CARRIED on straight, past these Indian suburbs of low-built huts and half-finished shacks to the Spanish center of the city, the *traza*, or grid plan. This was the exclusive home of those who held the rank of citizen and home too to their servants and household slaves. No formal barrier, no fortification or fence, marked the boundary between the Indian and Spanish cities, but a cultural and political gulf separated the two communities: the one subjected, the other isolated in its center. Each had its own laws and customs, its own jurisdiction, its own economy.

The division was plain to see in the ordered rectitude of the *traza* streets, which seemed like a chessboard to one observer and was the feature of the city that European visitors most frequently chose to describe in their accounts of the place. The characteristic grid plan of American cities remains strange to European eyes even today.

In 1555, an Englishman, Robert Tomson, explained that "the said City of Mexico has the streets made very broad, and right, [so] that a man being in the high place, at the one end of the street, may see at the least a good mile forward." His fellow countryman, John Chilton, who visited in 1568–1570, also marveled at the layout and construction of the place, with its "good and costly houses . . . built all of lime and stone, and seven streets in length, and seven in breadth, with rivers running through every second street, by which they bring their provisions in canoes."

This rectilinear regularity was considered absolutely essential to the civic structure. In 1531, a member of the city council commented on the importance of the *traza* and of ensuring that all the streets remained straight, and the following year the city officials agreed that the facades of buildings that did not precisely conform to the plan should be torn down.

The hustle and bustle of the cosmopolitan streets may have reminded the four survivors of the great Iberian ports of Seville and Lisbon, where all creeds and colors rubbed shoulders in the pursuit of international trade, but straight streets with the facades of all the buildings lined up like soldiers on parade were a novelty.

Although the streets were not paved with gold, the city had recently embarked on a scheme to improve the paving. Until that time, only the main thoroughfare, the calle Tacuba, was surfaced. Municipal improvement projects were continually thwarted as the building blocks and paving stones for these public works went missing, stolen for the construction of private houses by builders desperate for materials.

Calle Tacuba ran straight between the regular ranks of the conquistadors' residences, imposing palaces and town houses built of the dark purple volcanic stone of Mexico. With their carved baroque doorways, a strange hybrid of European and Aztec decoration, these homes were eloquent testimony to the prideful Spaniards' deeds of conquest. While along the center of the street a great aqueduct carried a constant flow of fresh water from the forests of Chapultepec, along each flank market vendors hawked their wares and no doubt called out to the famous four.

Although no physical barrier isolated the Spanish *traza* from the surrounding Indian city, the houses, monasteries, and churches were well buttressed; their windows were protected by solid iron bars and their doors were thick and hefty and armored with large nails. The castellations were no mere adornments, but designed for defense.

Mexico was a place of excessive wealth and opulence. Such was the ostentation of Spanish Mexico City that a royal decree was sent to the viceroy banning the import of luxury textiles such as brocade and golden cloth, in a futile attempt to stamp out the vulgar displays that sometimes seemed to be the very purpose of life itself. Such vanity was the daily business of the many Spaniards, their wives, their children, and their mistresses who had gorged their appetite for materialism on Montezuma's gold. But the viceroy, Antonio de Mendoza, was a shrewd governor who realized that it would be sensible to delay the public announcement of that royal decree. Instead, he suggested that in such an unruly colonial capital it would be more prudent to persuade the citizens to wait until their current gilded, bejeweled garments had worn out. He tried to curtail the excess of vulgar competition through quiet warnings about the imminent ban and urged the offenders not to replace their ludicrous finery.

But this was not to be, for wealth was at the heart of the matter, and the streets of Mexico were all but paved with gold. America was a land where humble peasant Spaniards might live like their sovereign and, what is more, do so without the cares of state that wore him down. They banqueted like the kings of Christendom and dressed the part; and like the pharaohs, Cleopatra, and the Caesars they also dressed their slaves with pomp and finery.

Thomas Gage, an Englishman who traveled to Mexico in 1625, was astonished to find that:

The gallants of this city show themselves, some on horseback, and most in coaches, daily about four of the clock in the afternoon in a pleasant shady field called la Alameda . . . where do meet—as constantly as the merchants upon our exchange—about two thousand coaches, full of gallants, ladies, and citizens, to see and to be seen, to court and to be courted. The gentlemen have their train of blackamoor slaves, some a dozen, some half a dozen, waiting on them, in brave and gallant liveries, heavy with gold and sliver lace, with silk stockings on their black legs and roses on their feet, and swords by their sides. The ladies also carry their train by their coach's side of such let-like damsels . . . who with their bravery and white *mantillas* veils over them seem to be, as the Spaniards saith, "mosca en leche," a fly in the milk.

Esteban soon saw that in Mexico, as in Spain, African slaves were a status symbol to be paraded in gorgeous finery. They were a direct demonstration of wealth, but they could also be titillating possessions who promised the almost Oriental eroticism of an exotic courtesan, like Doña Nufla and Doña Zangamanga, who set the Spanish court

alight with their beauty and licentious charms. Black slaves were to
be used and abused, and if most were forced to serve as beasts of burden, the luckiest were mere household pets, preened lapdogs amid the
glitter and pretension of gaudy aristocrats and powerful parvenu peasants. Esteban was well used to God's whimsy and the vicissitudes of
life, but even he must have been astonished by this gross jewel in
Spain's American crown.

Esteban also knew that for all their finery, most of these living
dolls had experienced no freedom since they had been so untimely
ripped from the lands of their birth. He too had lived as a slave, but he
had since known a kind of freedom as he crossed America. Yet he had
refused the chance to stay among his Indian followers and remain a
revered shaman. Instead, he had come to Mexico. What kind of liberty was he searching for?

4

FIESTA

July 1536

CABEZA DE VACA, Dorantes, Castillo, and Esteban were received as celebrities and treated as the honored guests of the most important citizens. They arrived in Mexico City on a Sunday, as the town prepared for the fiesta to mark the holy day of James the Great, Santiago, the patron saint of Spain. There were bullfights, and riders competed at tent-pegging, the bellicose sports of a military citizenry.

The survivors themselves became part of this pageant. Years later, Alonso de la Barrera, who had escaped the Narváez expedition because he had stayed with the ships, recalled seeing the four survivors at the cathedral in Mexico City. They were, he remembered, wearing buckskin or animal pelts (the Spanish is not precise) and looked like Indians, dressed "as they had been when they arrived from Florida."

As so often with the fragmented accounts of history available in the documents, the nature of the scene he is describing is uncertain. But evidently the Spanish authorities had arranged for the four men to again dress in their Indian clothes as part of the spectacle, or perhaps they had never changed their dress at all, as *Shipwrecks* implies. No doubt this was some kind of public exhibition of thanksgiving and celebration, an opportunity for God's merciful deliverance of the four men to be demonstrated to fascinated spectators already titillated by strange rumors of shamanism.

Dressed as Indians, the four men no doubt took communion in the cathedral, having confessed their sins. As they felt God's grace channeled through the hands of Archbishop Juan de Zumárraga himself, the important men of the city watched in wonder. The survivors had already become players on the political stage, pawns perhaps, but characters in a struggle for the control of information.

The Spanish settlers in Mexico had learned that the Aztecs shared the European love of ceremony and civic pageant; they had also learned the value of public fiestas in Indian society. For many years, the missionary friars had been aware that the Indians were greatly skilled in reconstructing natural scenes as backdrops against which to act out Aztec rituals, and the missionaries had made use of those skills in their own passion plays and liturgical dramas.

One missionary describes how at Easter, the story of the fall of man was acted out against a stage set made by Indian craftsmen which re-created high crags and mountains overrun by wild animals. "Adam and Eve's dwelling was so cleverly done that it did indeed seem like the earthly paradise, with its various trees and fruits and flowers, some of them real, some of them made with feathers and gold." The trees were "filled with all manner of fowl, with owls and carrion-eaters, and tiny birds and many parrots" and "the squawking and screaming [were] such that sometimes the play had to stop."

During the Corpus Christi processions, the streets along the route were so lavishly decorated that those who saw them claimed that if they were to describe it all back home in Spain, people would simply say they were mad. Fragrant grasses were strewn on the ground where the holy sacrament was to be carried. The route was marked by more than 1,400 triumphal archways, each covered with flowers and roses. The procession halted at six specially constructed altarpieces.

But these festivities were combined with more raucous revelry, which offended Zumárraga's rigid sense of decorum. He complained bitterly that it was "utterly contemptuous and shameless" for men to dress up as women and to dance and jump about in front of the sacrament. These dancers, he moaned, moved with lascivious sensuality and caused such a racket that they drowned out the religious chants. He was especially riled by the appearance of an effigy of the god of love, "so hideous to the presence of Our Lord," and thought the Mexicans would believe such profanity was part of divine law or doctrine because their own traditions celebrated their pagan idols with similar partying.

Bernal Díaz del Castillo, who had participated in Cortés's conquest of Mexico, described the banqueting that took place during special fiestas organized in February 1539 with the true pride of a founding father. Mexican ingenuity for organizing such events, he stated, was unrivaled in all the world.

"We awoke to find the main square filled with a forest of trees," so realistically arranged that they looked as though they had grown there. In the middle there were rotten trunks of trees that seemed to have fallen due to age. Some were covered in moss, with plants growing out of them. From the boughs of yet others, creepers hung like a veil. Deer, rabbits, hares, foxes, and all sorts of small beasts gamboled about these wonderful woods, and two little lions and four small tigers were kept corralled, ready for the hunt. And above, the forest canopy fluttered and chirruped with the sounds of a thousand birds, such a range of species native to New Spain that it would be impossible to remember them all.

Set apart from the main area of forest there were thick coppices, in each of which a squadron of savages lurked, imprisoned in corrals, armed with garrotes and bows and arrows. Suddenly, they were set free and rushed headlong into the chase, pursuing the game across the temporary hunting ground of the main square. Then, in the midst of the sport, two groups of these savages fell to fighting among themselves, settling some difference there must have been between them.

Then, the African cavalry processed across the square, forming a guard of honor for their "King" and "Queen." There were more than fifty in that regal train, Díaz remembered, all dressed in magnificent splendor, each man and woman adorned with gold and precious stones, sparkling in their pearls and silver jewelry.

When the following day dawned, the city of Rhodes had appeared in the middle of the Plaza Mayor, with its towers, battlements, its castellations, machelations, and ditches. A hundred gentlemen landowners rode onto this living stage, with their lances and bucklers, while other soldiers came armed with harquebuses or crossbows. The role of Grand Captain and Master of Rhodes was played by Cortés himself. He arrived with five ships, each with its masts and sails. The ships fired thunderous barrage after thunderous barrage and their trumpeters blasted out voluntary after voluntary, as they "sailed" three times about the square. It all seemed quite astonishingly real.

These spectacular charades were watched by the ladies and *señoritas* of the city, wives and consorts of the conquistadors. They graced the richly decorated windows of the grand buildings that gave onto the great square, dressed in fine bejeweled silks and damask. Young gallants served the women, bringing them sweets of almond, marzipan, and crystallized fruits, each decorated with gold or silver leaf, some bearing the viceroy's coat of arms and others the blazon of the marquis Cortés. These sweets were served with the best wines and the bitter chocolate drink of the Aztecs.

On consecutive nights, two great banquets were thrown, one by Mendoza and the other by Cortés. Mendoza had the corridors of the viceregal palace decorated with fruit trees and flowers, a verdant world of indoor bowers, filled with songbirds. Water poured from a fountain that made the centerpiece of this Arcadian scene, where a tiger was tethered, while its companion, a giant of a man who was dressed as a teamster, served drinks from two great wineskins.

Díaz tried to record the menu for posterity, although by his own admission he could not remember every dish. There were three or four types of salad, followed by roast kid and shoulders of ham, game pies filled with quail and pigeon, chicken blancmange, slow-roasted rooster, stuffed hens, turkey, partridge, and almond cake. After a break, the tablecloths were changed and basins and towels were brought so that the guests could wash their hands. Many of the dishes went untouched, but still the kitchens continued to send forth food: fish and turkeys that were boiled whole, with their beaks, combs, and feet gilded and silvered. For the sake of display, the heads of the deer, the pigs, and the beef cattle were cooked and brought to the table. All the while musicians sang and played. There were harps and violins, flutes and drums; and whenever the wine was served, the pipers struck up some appropriate tune.

In 1536, similar feasting marked the fiestas of Saint James, the warrior saint the Spaniards called "the slayer of Moors and Indians." Esteban was a guest at these valiant revels. Never before was an African slave so celebrated on such a sumptuous occasion. But the pomp surrounding the Spanish feasting was a shocking contrast to the simple celebrations of the Indians who had feted the four survivors during their journey. To the four survivors, it emphasized their own material poverty in comparison with the wealth of their spendthrift Spanish hosts.

Yet by a different reckoning they were wealthy men, rich in experience and knowledge. Their glory and their influence lay in their story, and their power lay in the way they told that story. Knowledge, experience, and power were as much Esteban's as his companions'. Esteban too could tell the story and there were many powerful men who were eager to listen.

It seems likely that all four men at first hoped they would return to the peaceful lands to the north with a royal mandate to rule the Indians who had revered them. They would have begun to dream of becoming rich landowners, princes of an American paradise, free to trade the riches of their fiefdoms with the rest of Mexico and New Spain and live as lords.

But any such dreams were forged of fantasy. Politically, Spanish Mexico was a place where rival factions fought vicious verbal battles as they vied for control of power and wealth. When the four survivors arrived in the summer of 1536, trouble was already brewing over the rights to explore the north. The four men now became pawns and players in these power games, and Esteban was to emerge out of that unpredictable political situation as the unlikely leader of the next northern expedition. We must assume, therefore, that he told his version of the story well, but inevitably it was never written down.

Even as the carousing of the Spanish fiesta continued, the political intrigue began. As the wine and the mezcal or pulque flowed at interminable banquets the four were lobbied by rich, powerful citizens, as men and women gathered to hear their stories, others fought for control of those stories. Alliances might be made, clues interpreted, and information gleaned amid the festivities. Wine and women loosened tongues. All imaginable forms of Machiavellian diplomacy must have been deployed by the many would-be intermediaries.

Among these powerful factions were the supporters of Hernán Cortés, the conqueror of Mexico, recently returned from Spain with the title Marquis of the Valley of Oaxaca, the greatest aristocratic title in Mexico. But it was an empty triumph for Cortés. Charles V had curbed much of his power and authority in exchange for that title, and Cortés would prove no match for his most cunning rival, the viceroy, Antonio de Mendoza, direct representative of the Crown of Castile.

Mendoza was a measured man, Charles's solution to the over-mighty Cortés. His job was to shore up the political life of the Spanish colonial capital and subdue the factions. But Mendoza's political position was complicated. An outsider, he had not conquered the territory himself as Cortés had done. He had made no stealthy alliances, nor sealed personal ties with brutal military tactics. He had never set his dogs of war on helpless unarmed Indians to strike terror among the people he aimed to vanquish. He had not been at Cholula, where Cortés's men slaughtered 3,000 Indian allies overnight when whispers of betrayal swept through the camp. Nor had he tortured the sacred Aztec priests and untouchable caciques to force out of them more gold and information about where it came from. Mendoza was unproved as a conquistador.

Mendoza's uneasy peace with Cortés was anchored by their common cause against Nuño de Guzmán. Both coveted the right to explore the northwest, but their path was blocked by the rogue colony of New Galicia. The four survivors were useful witnesses to Guzmán's many crimes and to his incompetence, and it may be no coincidence that within a year of their arrival he was jailed in Mexico. Guzmán is finished, Mendoza must have told them. Justice is closing in on him. There will soon be another governor in New Galicia. Then, we will arrange the peaceful, evangelical conquest of the north and you will lead that expedition. So, spare nothing when you report to the Emperor on what you saw in New Galicia, and I will see to it that your valiant Christian deeds are proclaimed as examples to us all. With words such as these the viceroy put pressure on his valuable but vulnerable witnesses.

With the four survivors in their service, Mendoza and Cortés could act against Guzmán. The Spanish monarchs had always been concerned about their moral Christian duty toward their new Indian subjects, because the mortal souls of these barbarous and heathen peoples were the responsibility of the sovereign, by divine law. Their conversion and salvation were crucial to Charles's own chance of gaining a place in paradise when the Last Judgment came. How best to kindle true faith in the native population of America was a corporate responsibility of Christendom. God would brook no failure in such a sacred task—Charles could be sure of that.

The Crown was well aware of these problems. In 1526, Charles V issued a decree outlining the injustices that had been reported to him:

> We have learned of the ungoverned greed of some our subjects who sailed to our islands and mainland in the Ocean Sea, and the great and excessive hardships forced upon the Indians, in the gold mines and in the pearl fisheries, and in agricultural work, where they are made to work excessively, without moderation. They are given neither the food nor clothing necessary to keep themselves alive and are treated without love and with cruelty, much worse than if they were slaves. All this has caused a great number of them to die, so many that some places on the islands and the mainland are now barren and without any native population. Others have left their own lands and natural habitats and are gone to the scrublands and other places in order to save themselves and escape their subjugation and ill-treatment. This has been a great obstacle to the conversion of the said Indians to our Holy Catholic Faith.

Charles ruled that such ill-treatment should stop. But Guzmán's New Galicia depended on trading Indian slaves, and that, as Cortés and Mendoza knew, could be Guzmán's undoing.

We have no description of the sale of the consignment of slaves who had been taken by Alcaraz and Cebreros and who had traveled alongside the four survivors all the way from Culiacán. But we do have a disturbing description of the sale of New Galician slaves in Mexico the year before, 1535, written by two magistrates and a notary of the Mexican Audiencia.

Men, women, and children, even infants as young as three or four months old, were all branded on the face with the royal brand, using a hot iron that was the size of the children's faces. "We saw all this and more. . . . Some of these people were so ill that they were almost about to die," they noted. They asked the owners, "What evil have these women and suckling babes done to deserve such branding?" And the owners replied that they had brought these people down out of the mountains by force because they had fled from the Spanish soldiers.

Just as sheep flee from wolves, so the Indians flee, escaping from innumerable wrongs, injustices, and injuries, the magistrates explained. These people, who are as mild as rabbits, defend themselves with the only weapon they know. They flee into the bush and scrubland while we, blind to our own greed, determine that such a natural reaction shall be deemed rebellion because it suits our own private gain to do so.

These people, they complained, were in fact treated worse than slaves. They were sent like convicts to work in the mines, their faces "plowed up" from having the brand or name of every owner burned into them. "Thanks to our sins, some bear the marks of three or four different brands, so that a human face, which was created in the image of God, is used as sheet of paper, not by fools, but worse still by men of greed."

With such scenes as these taking place outside the windows of Mendoza's palace, it must have been easy for the canny viceroy to persuade the four survivors to spare no effort in remembering every detail of those frontier crimes committed brazenly by Alcaraz and Cebreros.

Although Guzmán's brutal treatment of Indians was well known, there were many who argued that the savage Indians had no souls, or that they might be taken as slaves in "just wars." Even Charles V accepted that it was better to enslave Indians than to have Spaniards abandon their colonies.

But if three impoverished Spaniards and a black slave, traveling "naked and barefoot," could harness the collective power of many different Indian tribes, bring peace among them, and preach Christianity to them, then surely Guzmán's gangsters now had no excuse?

This argument was taken up by the awe-inspiring and authoritarian figure of Archbishop Zumárraga, a man close to Mendoza, a mortal enemy of Guzmán, and the person who perhaps held most influence over the four survivors when they first arrived in Mexico. A tough churchman from the Basque country who had first led the opposition against the disastrous and divisive government of Guzmán, in 1529, Zumárraga had forced his enemy out of Mexico, and like Cortés and Mendoza he was anxious to see Guzmán brought to book.

JUAN DE ZUMÁRRAGA'S career had begun humbly enough, in the villages and farmsteads of northern Spain. There, he devoted himself to the parochial pursuit of sorcery in the semi-pagan uplands around Pamplona. He was a ready Inquisitor and an avowed conqueror of witchcraft, the occult, and all other demonism and devil worship. Quite unlike Cortés or Mendoza, he was motivated by a fearsome zeal, not money. Power, for Zumárraga, lay in winning souls for the church through the successful conversion of the pagan Indians. To such a man, the four survivors and their story were a godsend.

He was a Franciscan friar, fanatically devoted to rejecting all that was worldly and material in favor of a deeply spiritual life. Saint Francis had preached that the heathen were to be converted by example; his followers were to live by hard work, fasting, and begging. Franciscans believed fervently in their own poverty and in denying themselves all comforts and luxuries. The idea that they should reject all but the most basic of their own bodily desires was absolutely central to their understanding of religion. It was their spiritual, moral, and social duty to model their lives on Saint Francis, who had modeled his own life on Jesus Christ. Franciscans also believed that the history of the world was divided into three eras: that of God the Father, described in the Old Testament; that of God the Son, described in the New Testament; and the millennium, the era of the Holy Ghost, which would see the end of the church and the rise of pure spirituality. The Franciscans quickly adopted America as the model for that marvelous millennial land.

Zumárraga was one of the key religious figures in the New World who rejected the established Spanish military conquest of the Indians in favor of peaceful evangelization. In 1536, he was providing support for his friend and fellow Franciscan, Vasco de Quiroga, who was attempting to establish a peaceful mission near Guadalajara. Both men were heavily influenced by Bartolomé de Las Casas, the most famous theologian to preach peace both in the New World and at the Spanish court. As a young man Las Casas was involved with Pánfilo Narváez's infamously barbaric campaign of subjugation against the native population of Cuba, which preceded his ill-fated expedition to Florida—the expedition that Cabeza de Vaca, Esteban, Dorantes, and Castillo had so miraculously survived.

According to Las Casas, Narváez exuded authority and self-confidence. Tall and blond, he had a way with words, and was very much the perfect image of a gentleman. But he was "not much given to caution," the priest explained, and he loved to do battle with the Indians. In 1511, he arrived in Cuba leading a company of archers, eager to indulge his love of hunting human beings. The governor, Diego Velázquez, welcomed him, giving him free rein to proceed with the pacification of the island.

One story told about Narváez by Las Casas, which has been much quoted, provides a plain perspective, across the years, of the great differences between the ways in which Indians and Europeans approached each other during those first years of contact.

Narváez and a troop of twenty-five or thirty archers and infantrymen had billeted themselves on an Indian settlement while they set about subjecting the surrounding population. Their hosts at first seemed compliant enough, bringing regular supplies of food and silently suffering the unwelcome attentions paid by the lusty Spaniards to their women.

But the uneasy peace was broken in the dead of night. Narváez's sentries slept, lulled into complacency by their own arrogance and deceived by their hosts' dissembling. Their commander had failed to ensure due vigilance and was rudely woken from his slumber by the war cry of 7,000 braves descending on his camp.

The Indian leaders had divided their forces so as to attack the Spaniards on both flanks, but the innocence of their motives and tactics in this attack seems extraordinary. This innocence was the undoing of the Indians and makes us realize that for men like Narváez, aficionados of such battles, the business of fighting Indians was more akin to hunting than to war.

The Indians fell upon the camp and burst into the huts where the Spaniards slept, "but they neither killed them nor injured them, but instead they grabbed up their clothing, which was the only purpose of their attack." They had come, en masse, to steal the Spaniards' clothes. Only Narváez himself was subject to any physical assault. As he rose sleepily from his bed, one of the raiding party threw a stone, which hit him in the stomach and winded him. But he had his horse swiftly saddled and, dressed only in his nightshirt and carrying a string of bells, he rode unarmed through the camp. This bizarre, harmless, quixotic figure proved so terrifying to the Indians that they fled back into the forest. Such was the threat the natives of Cuba posed to their would-be masters.

Las Casas repeatedly saw the bloodlust of these conquistadors during the two years he witnessed Narváez's troops progress from peaceful village to peaceful village. The dismayed friar described how at one such settlement, one of the soldiers, "possessed by the devil," drew his sword completely without warning as a signal to his companions.

All the Spaniards now unsheathed their weapons and began to slash at the Indians, "eviscerating and murdering that flock of sheep and goats." None were spared; "men, women, children, and the old, who were sitting around unsuspecting," were cut down. The murderers now entered a great lodge, where they attacked the defenseless Indians who cowered inside, chopping and thrusting, opening up their victims, cutting men in half with the swords they had assiduously sharpened that morning on the boulders where they had crossed a beautiful, babbling brook.

Las Casas managed to prevent the massacre of the expedition's forty Jamaican porters. But their five guards, faced with his objections, abandoned their posts and joined their comrades in the slaughter of the Cubans going on outside. Las Casas followed them and remonstrated with Narváez, who was watching impassively.

"Good sir," Narváez mocked, "what do you think of our Spaniards and their deeds?"

Las Casas ignored the gibe and did his best to persuade the Spaniards to stop. And when their bloodlust seemed satisfied and they stood panting from their efforts, their weapons blunted by the breaking of so many bones, he entered the great lodge and tried to reassure the Indians that his countrymen were finished with their sport.

But no sooner had one young man climbed down from the rafters than he was cut down by a soldier whose appetite was as yet unsatiated. The Indian stared down at his own entrails, hanging from his belly, and collected them up as best he could. He staggered to Las Casas and begged him for baptism.

It seems that Narváez learned from this Indian's innocent belief in Las Casas's authority. On another occasion, he let the priest send messengers into the Cuban hinterland, calling on the Indian leaders to agree on a peace. But when the twenty-one chiefs appeared with their retinue, Narváez broke his promise and made preparations to burn them alive for their perceived insurrections.

Las Casas was so horrified by the wanton violence that he returned to Spain to become a Domincian friar and soon became an outspoken critic of the Spanish colonists, publishing a damning indictment of their brutality, *A Brief History of the Destruction of the Indies*. He has become known as the "Defender of the Indians." Instead of armies, Las Casas argued, Spain should send Christian soldiers to convert the heathen.

By 1536, Las Casas's protests had helped to engender widespread condemnation of the conquistadors by Spanish theologians. That year, in an aggressive letter to Viceroy Antonio de Mendoza, Archbishop Zumárraga railed against the Spanish government in America. He demanded that more friars and monks be sent to America so that they could extend their missions into new territories. "That," he explained, "is why I asked and pleaded with the Council of the Indies to be allowed to name thirty friars." But instead, he had arrived understaffed because the council told him that twelve would be plenty. He regretted, he said, "that he had not borrowed more money himself so that he could bring those thirty pious men, for it would have pleased God."

Las Casas gleaned useful ammunition for his battery of books and words from another Franciscan friar, Marcos de Niza, who claimed to have witnessed Spanish atrocities committed during the conquest of Peru. In the summer of 1536, Marcos de Niza was in Mexico at Zumárraga's invitation.

For these religious men of peace, the account given by Cabeza de Vaca, Esteban, Dorantes, and Castillo clearly had enormous potential. Finally, they had exceptional evidence that, when appropriately manipulated and presented, would prove to a skeptical world that a few good men could peacefully bring thousands of Indians into the Christian fold.

Zumárraga at once set about forcing the four survivors to give an account that would suit his purpose. They themselves were no doubt sympathetic to his cause anyway, but in case they dissented, he had a powerful means of controlling his "pilgrims," as he soon called them. As well as being the first Archbishop of Mexico, Zumárraga was also the first Inquisitor General.

Since arriving in Mexico, Zumárraga had become ever more concerned by the apparent Satanism of Indian culture. It seemed to him that the devil's work bubbled below the surface of Mexican Christianity, much as the Jews in Spain had continued to practice their religion after they were baptized, and Zumárraga could be a brutal Inquisitor. In fact, he was officially reprimanded for his severity in ordering that a baptized Indian chief be burned alive for heresy in 1539.

The four survivors therefore had every reason to be worried as they waited for their first meeting with this ferocious Franciscan. The strange ad hoc alchemy of shamanism, faith healing, and Christianity out of which they had concocted their godlike status among their Indian companions was potentially heresy as far as the Inquisition was concerned. They had little doubt that Zumárraga could easily view such heterodoxy as worthy of severe punishment. To stray from the path of normal Catholic Christianity in early modern Spain or Mexico was a dangerous business, and all four survivors, when they were still in Spain, had seen heretics burned at the stake.

With such a powerful threat in his armory of argument, Zumárraga could be quite sure that the four survivors would do his bidding and present their strange experience as a divine Christian miracle. We do not know what passed between the Archbishop and the survivors, but we do know that the official account they gave—the version that was then written down and sent to Spain—was conspicuously convenient for Zumarrága's personal political cause. The survivors' story dwells disproportionately on descriptions of the social disasters caused by Guzmán among the Indians of New Galicia. Nor did Zumárraga miss the opportunity to discredit Pánfilo Nárvaez, the butcher of Cuba, so hated by Las Casas. When the four survivors came to the

Audiencia of Mexico to tell their tale to Viceroy Mendoza and the city magistrates, they destroyed Narváez's reputation and ridiculed his leadership.

But if Zumárraga was determined not to miss this opportunity to thoroughly discredit Nuño de Guzmán and Pánfilo Narváez, his principal purpose was to affirm the policy of peaceful evangelization preached by the coterie of priests who sought power through righteous words and pious action, rather than with the sword.

And so he had set about changing his pilgrims' story of extraordinary survival, made possible by their command of native shamanism and heathen religion and the heterodox wisdom of an African slave. For those actions he might have had them burned at the stake to save their souls from his terrible God. Instead he chose to turn the story of their experiences into a parable of Christian suffering and redemption through miraculous divine intervention. This was to be a story worthy of Saint Francis or Christ himself: the four survivors would be presented as disciples of Christ, or at least as modern saints blessed by God.

The survivors had no reason to object, for in Zumárraga they had a champion who seemed to promise them the opportunity to return to the Indian country in the north, but this time with a lifeline to the Spanish world.

Zumárraga now wrote for Mendoza a report on the treatment of Indians, parts of which were built around his interviews with the four survivors. He argued that instead of sending armies to conquer and explore the north, it would be better to send friars and monks to preach to those people, just as Christ had sent his apostles and disciples in peace. Little by little, these churchmen could penetrate the Indians' lands, their homes, and finally their hearts. Instead of attacking the Indians, they would build churches. And the four survivors became the centerpiece of his argument.

We can learn much, Zumárraga argued, from the account given of their experiences by the men who survived Narváez's expedition. They spent many years living among the Indians, traveling through remote lands where the gospel had never been heard and where people knew nothing of Christianity until these men preached to them. The Archbishop pointed out to his viceroy that contrary to the widespread misconception of the Indians as cannibals, they had treated the survivors much better than Christians would have been likely to do under the same circumstances. In fact, the survivors were so successful in developing good relations that the Indians came to revere them with the zeal that Christians reserve for their saints.

"I am quite persuaded," Zumárraga wrote Mendoza in 1536 or 1537, "that it has been proven beyond reasonable doubt that if a few friars whom I know here in New Spain were to go with these Narváez survivors to those lands, traveling in the same way that they did, walking among the Indians, and if they continued into other parts of that wide expanse of territory, they would show that war is unnecessary." This argument would lead directly to Esteban's appointment as the leader of a Mexican-Indian army that would eventually march north under his command, back into Arizona and New Mexico. And Zumárraga proposed that his new Franciscan favorite, Marcos de Niza, should be the spiritual leader in that endeavor.

Zumárraga and Niza seem to have begun to believe their own rhetoric. It was the same mistake that had undone Narváez and many other conquistadors, whose arrogance made them blind to their own ignorance about America. For these fanatical Franciscans, the whole business seems to have quickly achieved almost biblical significance, thanks to their willingness to believe and promulgate a much manipulated version of the survivors' story as proof of their own theories.

In hindsight, the plan for the conquest of the north as proposed by Zumárraga is manifestly absurd. His intention was to try to gain control over almost all of what is now the United States and Canada with a handful of monks and friars and the four Narváez survivors. But in an age as medieval as it was modern, Europeans sometimes had little grasp of reality, while their intense faith made them believe whatever they might want to believe. Zumarrága had wanted a Christian story of peaceful evangelization through personal suffering and the power of faith and prayer. Whether it was true or false was not important. All that mattered was that it should be believed. In the survivors' tale he got what he wanted, a perfect mixture of fact and fiction. But it led to abject failure for Marcos de Niza and his peaceful mission. It also led to Esteban's final freedom, once more among the Indians.

And so the story told by the survivors to the Mexican Audiencia and later embellished seems today like something taken from the early history of Christianity. It is filled with accounts of suffering and hardship that lead to redemption and the conversion of pagans. Even the practical governor of Venezuela—a German and a banker—was happy to describe the survivors as "apostles" when he first heard of their adventures. More recently, some scholars have referred to Cabeza de Vaca as a "messiah." A friend of mine from Seville very kindly let me see an early draft of a forthcoming paper in which he draws analogies with the story of the prodigal son. Another scholar points out that Las Casas could not have nailed his colors to the mast of a story that more perfectly fitted his own rhetorical position.

Zumárraga's role in shaping the story has hitherto gone unremarked because it is impossible to prove his involvement conclusively. But Zumárraga had both motive and opportunity—which is not to say that the four survivors who had so manifestly failed in their military conquests were in any way reluctant to wear the mantle of Christian heroes.

ANTONIO DE MENDOZA treated the four survivors as his personal guests in the viceroy's palace. Years later, Dorantes's stepson remembered that he had entertained them lavishly and had given them all of the gentlemen's accoutrements they needed: clothes, horses, and weapons. But the price of this luxury was information, and he pressed them for a private account of the north. They told him about myriad walled market towns in the far-off provinces of the upper Rio Grande, great metropolises with buildings six or seven stories tall. The four survivors, it seems, were quite as capable as Zumárraga of embellishing their own story to suit their own purposes.

Mendoza's viceregal palace was the center of colonial life and government in Mexico City. It dominated the central square, heavily built and castle- like, a bastion of colonial strength at the heart of the town. It was crowded on all sides by the bustle of the marketplace, and the ground floor was filled with artisans and craftsmen. Their workshops offered a loud, cacophonic backdrop to the crowds of Spaniards who went hither and thither, some on foot and some on horseback. Inside there was the central post office, with many scribes and notaries ready to write letters of business, romance, or mundane family matters for the illiterate.

On the first floor crowds of nervous fellows whispered to one another, murmuring in low tones until an occasional voice was raised in anger. Their business was the legal world of the Audiencia; they were litigants and defendants, soldiers, merchants, and citizens, and their solicitors and advocates. In the Audiencia itself there was a reverent silence. The viceroy, magistrates or justices, and other officials viewed the proceedings with solemn gravitas.

Mendoza and Zumárraga had prepared the four survivors to make their official report to the Audiencia, where the viceroy and a body of magistrates would hear their testimony, while a notary recorded the story. That official process of legally validating their report produced the first written account of their adventures.

The first and most important testimony was given by Cabeza de Vaca, the treasurer of the Narváez expedition, whose job it was to account for what had happened. He took the stand for many hours—

spread, no doubt, across many days—recounting in great detail the collapse of the expedition and the loss of almost all the men.

In due course, Cabeza de Vaca reached a point in his story where he recounted his separation from the few survivors who remained alive in the spring of 1529. The officials were little interested in what had happened to him for the following five years, during which he lived alone among the Indians, so they now turned to the next most senior witness, Andrés Dorantes, to hear the fate of the main body of the expedition. Dorantes too took the stand for many hours, describing in detail an unfolding tragedy.

From time to time, the viceroy and magistrates no doubt turned to Castillo and perhaps Esteban for corroboration, but mostly they listened as Dorantes kept them spellbound with his tale.

The original manuscript of this hearing is now lost. But Cabeza de Vaca and Dorantes kept copies, and each embellished and elaborated these copies until they had created two versions of the story. These two versions have survived: Cabeza de Vaca's *Shipwrecks* was published in Spain in 1542; Dorantes's account became the basis of the official history of the expedition written by the royal historian, Gonzalo Fernández de Oviedo.

But, before examining that story of Pánfilo Narváez's failed expedition to Florida and the first crossing of America by non-Indians, I will turn next to Africa and then Spain in order to understand how Esteban came to be involved in this epic adventure.

Part Two

ESTEBAN

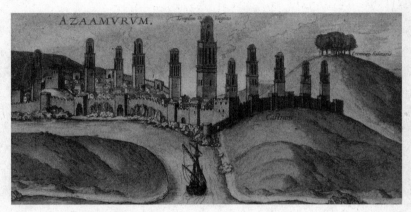

Azemmour, from Georg Braun and Franz Hogenberg, *Civitates orbis terrarum* (1572). (Courtesy of the Jewish National and University Library, Shapell Family Digitalization Project and the Hebrew University of Jerusalem, Department of Geography, Historic Cities Project)

5

"NEGRO ALÁRABE"

WHO WAS ESTEBAN? Where did he come from? What was his background? How was he raised? What youthful experiences formed him?

It is impossible to answer these questions with precision because Esteban was an African slave. He almost certainly came from one of the west African kingdoms to the south of the Sahara desert, but no birth certificate exists; no entry in the parish register has survived. We know that he lived for a time at Azemmour, today a small, unimportant town on the Atlantic coast of Morocco, but there is no documentary record of his life there to tell us what he did or even who his owners were.

The fact that we have little concrete information about Esteban's youth makes him emblematic of the wider African-American experience. The modern social concept of an African-American assumes ancestry in Africa and a story in which the point of origin, the seminal event, is the export of millions of slaves to America by Europeans. But the idea of calling those slaves "Africans" was originally European. The slaves themselves would have seen their origins in terms of their family, tribe, or clan, and in the principalities and kingdoms with which they identified. They thought of themselves as Wolof, Mandinga, or Songhai; they were from Guinea or Mali. They had no concept of being "African"; nor would they have regarded themselves as "black" until they reached the European world.

For modern African-Americans, the intervening centuries of marriage and shared experience mean that precise African origins have become obscured and often uncertain. African-American history is colonial and American, underpinned by an imprecise sense of a prehistory in which Africa is an ill-defined, at times almost mythical place of origin. In the same way, Esteban's African story is obscure, and our sense of his historical identity begins when he reaches America and takes on a role in the Spanish colonial world. He is not the first African-American simply in a practical sense, but also symbolically.

However, although there is poetry in dwelling on the obscurity of Esteban's origins, it is possible to throw some light on what his early life must have been like. To do that, it is best to begin with the most specific evidence available. We know that *Shipwrecks* described Esteban as a *negro alárabe, natural de Azamor,* "an Arabized black, native to Azemmour." The apparently simple Spanish phrase *negro alárabe* is profoundly emblematic of an important problem, which has obscured and confused our sense of Esteban's origins and identity because, from time to time, historians and other commentators have suggested, against all the evidence, that Esteban was not really a black African at all.

To set the record straight and to understand why this is important, it is useful to repeat the essence of Richard Wright's argument of 1902, quoted in the Introduction. Wright concluded that "the useful and noble deeds of the Negro companions of the Spanish conquerors" had not been properly recognized because historians tended to see the masters as being entitled to credit for the work of their slaves.

At the time, Wright was specifically concerned with the way the prolific late-nineteenth-century philosopher and historian John Fiske had represented Esteban in his wide-ranging two-volume history, *The Discovery of America*, published in 1892. Fiske's reputation has always been considered tainted by his racism and theories of Anglo-Saxon superiority. He had thoroughly irritated Wright by playing down Esteban's role in the "discovery" of New Mexico and Arizona and referring to him as "poor silly little Steve."

Wright himself had been born into slavery, but after exemplary military service in the Spanish-American War, he became an energetic supporter of African-Americans' rights and was highly influential politically. He rounded on Fiske, determined to show that New Mexico and Arizona had been "discovered" by a "Negro," and not by Esteban's companion Marcos de Niza. Using the cunning of a lawyer and the precision of a scholar to artfully repackage Fiske's words to suit his own purpose, Wright wrote:

Indeed it seems clear that a fair interpretation of the facts related in Dr. Fiske's work would warrant the conclusion that [and here Wright quoted Fiske himself] "a man [Esteban] who visited and sent back reports of a country," is more entitled to the honor of its actual discovery than [Marcos de Niza] who, according to Dr. Fiske's own statement, "from a hill, only got a Pisgah's sight of the glories of the country, and then returned with all possible haste"—without having set foot actually within the Cibolan settlements of New Mexico.

Wright was careful not to accuse Fiske or any other historian of overt racism, but instead argued that black Africans had not been properly represented because history was a product of a hierarchical society. His argument was not with the institution of history itself, but with the racial prejudices that continued to influence the perspectives of historians.

In 1940, another distinguished African-American scholar, Rayford Logan, argued in the new academic journal *Phylon* that overt racism and prejudice of individual historians had led Esteban to be incorrectly described as a North African "Moor," rather than as a "Negro," and had denied his "discovery" of New Mexico and Arizona. Logan wanted to set the record straight and establish once and for all that Esteban was a "Negro" and had been the first non-Indian "discoverer of the southwest." He set about that task in the light of his own experience of racism in America, and accordingly he was interested not simply in the truth about Esteban, but also in how that truth had been suppressed.

At the time Logan was writing, the American southwest was celebrating the 400th anniversary of the "discovery" of New Mexico by Francisco Vázquez de Coronado. In 1540, Coronado had presided over a sanguinary sojourn into Arizona, New Mexico, Colorado, Texas, and beyond. It was a venture frequently punctuated by bloody brutality perpetrated against largely friendly Indians. In the first place, Logan, like Major Wright, wanted to make it clear that, however you interpreted the historical documents, one thing was certain: Esteban had "discovered" New Mexico before Coronado and was quite possibly its "sole discoverer." And he clearly believed that racism was one reason why "the Southwest chose to commemorate in summer-long *fiestas* the *entrada* of Corando" instead of Esteban.

Logan was also troubled by the fact that those in charge of the celebrations "had made Estevanico a Moor rather than a Negro." In this Logan erred somewhat, for most of the publications produced as part of the commemorative festivities refer to Esteban as a "Negro." The notable exception was an essay by Cleve Hallenbeck, a meteorologist, whose hobby was the history of the southwest. He claimed that Cabeza de Vaca's "plain statement that [Esteban] was an Arab leaves no room for argument." This assertion was based on his interpretation of the phrase *negro alárabe, natural de Azamor*; and it is an assertion that, when I first read it, seemed to me to be totally wrong.

So I turned to a respected authority on the Spanish language, a former teacher of mine, Professor John Butt, who explained that *negro* should be translated as "black man" and *alárabe* as "Arabic-speaking," or perhaps "Arabized." Whoever wrote these words meant that Esteban was a black man who was, in some way that is left unspecified, culturally but not racially Arab. It certainly does not mean a black Arab, a Moor, a Berber, or any other race usually associated with North Africa. Esteban was almost certainly Negroid and of sub-Saharan ancestry.

But as I read more, I found that Hallenbeck's ignorance of Spanish had not deterred him from further inaccurate explanations of why he thought Esteban was a Moor or an Arab. He also noted that one Spanish conquistador who had met Esteban referred to him as *moreno,* "brown"——not black. Hallenbeck obviously had no idea that Spaniards used the phrase *los hermanos morenos* to refer to a Christian brotherhood that was founded in Seville specifically to provide social and spiritual support to black slaves and freedmen of sub-Saharan origin, a Christian brotherhood popularly known as *Los Negritos,* even to this day.

Given Hallenbeck's breathtaking arrogance when it came to holding forth on the niceties of the Spanish language, it is hardly surprising that he also informed his readers that in Cabeza de Vaca's time, Spaniards applied the word *negro* to "people of Hamitic and Malayan blood as well as to Negroes." Of course, he was in no way qualified to make such an assertion, but on this occasion he was not entirely mistaken. He was simply wrong about how the word *negro* was used and what it meant.

Rayford Logan had himself pointed out that the Spanish sources nearly always refer to Esteban as *el negro,* "the black," and that nearly all translators had interpreted this as meaning "Negro." He also realized that Cabeza de Vaca used the same word, *negro,* to describe the human cargo of a Portuguese slaving ship. But Logan never came to grips with the fact that Spaniards at that time did sometimes describe Arabs and Moors as *negro*. So what did sixteenth-century Spaniards really mean when they referred to a man as *negro*?

Negro is simply the Spanish word for "black," and as such it was used to describe the color of objects and people's moods. In that sense, it was no different from the meaning of the word "black" in English today. But Spaniards used the word *negro* about people in two ways: they might use it as an adjective to describe people's skin color or even their character, whether they were Spanish, Moorish, or African; but when they talked of *un negro*, or *una negra*, "a black" (male or female), then they meant what Wright and Logan understood by the word "Negro."

In the Middle Ages, all over Europe and the Mediterranean, from England in the north to the North African Muslim lands in the south, from Spain and Portugal in the west to the Ottoman empire in the east, the far-off, mysterious world beyond the Sahara Desert was known as the Land of the Blacks. What was meant by this is made obvious by a world map made in Catalonia in northeastern Spain in 1375, which shows the Kingdom of Melley (from which we get modern Mali), south of the Sahara Desert. The King of Melley is clearly shown with characteristic Negroid facial features and hair. Nearby, a short commentary explains who he is:

This Black Lord is called Musse Melley, Lord of the Blacks of Guinea. This King is the richest and the most noble Lord of all this area because of the abundance of gold found in his lands.

In fact, although the Catalan uses *negre*, "black," to describe both ruler and subjects, the translation offered in a sumptuous edition produced in the 1970s describes the king as "this Moorish ruler . . . lord of the negroes of Guinea." In other words, as late as 1978 *negre* was being translated as "Moor" when used of a king and "Negro" when used of a subject!

Nearby, to the west and in the desert, the map illustrates a man with a pinkish face, who is riding a camel, and has an encampment of tents pitched nearby. He is evidently intentionally portrayed as having a much lighter skin color and more Caucasian or Hamitic features than his "black" neighbor. He is probably a Tuareg, for the description tells us that "all this area is populated by people who muffle up so that one can only see their eyes. And they live in tents and ride camels." It is clear from the contrast between this man and the "black" Musse Melley that the mapmaker wanted to show them as belonging to two different races.

Writers also made this distinction. In the 1450s an Italian merchant, Alvise (or Cadamosto), described his exploration of the west coast of Africa. He wrote about *nigri* and *arabi*, clearly distinguishing between blacks and Arabs. In exactly the same way, a Portuguese history written at the beginning of the 1500s differentiated between the *mouros* and *negros* of the Mandinga region. Near the end of the 1500s, a Spaniard described the five regions of Africa as Egypt-Ethiopia, Barbary, Numidia, the desert regions called the Sahara, and "the land of the *negros*" beyond the desert. Throughout the sixteenth century there was a sense that the *negros* were different from North Africans and were from beyond the desert.

In fact, the difference between Moors and *negros* cannot be explained more clearly than by looking at a grant made in 1594 by the King of Spain to his subjects in the Canary Islands. He gave them the right to raid North Africa for slaves, "because," as the royal decree explained, "the *alárabes* [Arabs] of that land have many *esclavos negros* [black slaves] and moreover . . . there are other Moors who ransom themselves, paying with many *negros* [blacks]." So the reason for attacking Barbary North Africa was to capture or extort the black slaves belonging to the Arabs and Moors.

And the reason they wanted "black" slaves is made quite clear by a seventeenth-century Spanish lawyer, Cristóbal Suárez de Figueroa. "When it comes to talking about the slaves you can get today, well, they are either Turks and Berbers, or Blacks. Of these the first two kinds are usually untrustworthy, being great sinners and criminals. But the Blacks," he said, "are much better natured."

Here we may glimpse Esteban's story, for he was probably once one of these black slaves owned by a North African Arabic-speaker. And it is clear that men like Esteban, who were from the sub-Saharan "land of the blacks," were closely associated with slavery. But to better understand how *negro* was used in the specific context of slavery, it is helpful to look at Vicenta Cortés's classic study of the Spanish slave trade for the forty-year period immediately before Esteban first arrived in Spain.

Cortés toiled for a decade or more in the Spanish archives, carefully reading the records of thousands of transactions made by dealers at the slave market in Valencia. No one could be better qualified to judge the meaning of *negro* (in this case the Valencia cognate *negre*). As one reads her book, it quickly becomes clear that her examination of the documents had revealed that there were two basic categories of slave: "white" and "black."

The documents describe various origins for "white" slaves. Some are Moors; some are from Barbary; there are even a few Jews. Among these there is an occasional reference to *blancos oscuros*, "dark whites." In other words, in these official documents and legal contracts, there was resistance to using the word *negro* or *negre* to describe white slaves. By contrast, most examples of *negros* that Cortés came across referred to men and women imported in large numbers by the Portuguese from their fortresses at Arguin, Cape Verde, San Jorge da Mina, and San Thomé, the engine rooms of the sub-Saharan trade.

But in these documents there are examples of apparent anomalies that seem particularly relevant to our discussion of Esteban's origins.

For there is at least one reference to a *negra mora* (Moorish black woman); elsewhere, there is one to a *negro moro* (Moorish black), who is later described as *el negro* (the black) in contrast to two *moros* (Moors) who had been captured at the same time. Many years later, Cortés explained that these "so-called Moorish blacks [were] converts to Islam, [whose] geographic place of origin [was not stated in the documents], which leads us to think that such examples refer to blacks who arrived from north African kingdoms, but [originally] came from further south." Again, we glimpse something of Esteban's origins alongside the *negros moros*, because these Moorish blacks seem to have also been "Arabized." They too might have been referred to as *negros alárabes*, as Esteban was.

More recently, Aurelia Martín has researched this subject in the archives of Granada. As a result of many hours spent in the company of documents written in the fourteenth and fifteenth centuries, it became so obvious to her that *negro* in old Spanish was used to refer to black Africans of sub-Saharan origin that she coined the term *negro-africano* to cope with the problem of how to make the distinction in modern Spanish. That she had to do so reveals just how confusing it is to use colors in the description of races and, more controversially, also suggests the latent typological inadequacy of racial taxonomy in the modern world.

In fact, the word *negro* when applied to slaves clearly meant something other than "black" in the sense of color, just as "black" does today in modern English. This is obvious in the way *negros* were often described by sixteenth-century Spanish notaries as "light," "dark," or even "very dark." This tells us that the people who wrote descriptions of slaves as part of their work thought of the word *negro* not as meaning the color black but as describing a "race" characterized by a range of skin colors.

The plain and simple truth is that this distinction between slaves who were *negros*, sub-Saharan Africans, and those who were *blancos* seems to be universally accepted by scholars of the specific subject of Spanish slavery. It seems that Rayford Logan was right to describe Esteban as a "Negro." But scholars who study the history of slavery now generally agree that "race" itself does not exist in terms of biological difference and that the idea of "race" is therefore outmoded. The apparent differences between people, it is argued, are best understood in terms of social and cultural distinctions, not skin color or other physical characteristics. And so it is sensible to ask if we should really worry about whether Esteban was a "Negro," a *negroafricano*, an Arab, or a Moor. After all, might it not be racist to make that distinction?

It is possible to avoid this problem by arguing that Esteban's identity as an African-American is what is important, and that how African-Americans may be defined is not the business of this book. As long as the concept of an African-American is current and as long as African-American history is seen as beginning with enslavement in Africa, then Esteban is important because he is the first African-American. Arabs, Moors, and Berbers rarely enter into that story except as slavers and slave merchants—people who had bought and sold slaves from beyond the Sahara Desert long before the Portuguese, the Spanish, or the English. Arabs, Berbers, and Moors are not part of the African diaspora; they were part of the Mediterranean and European world that brought about the diaspora, even though a few may have ended up enslaved themselves. In fact, it is thought that as many sub-Saharans were forcibly removed from Africa along Islamic trade routes as were shipped to America by Europeans.

Esteban was born during a period in history when those who survived were all too often traumatized by life itself. It was an era that was not merely scarred by violence, disease, and famine but was also sickened by the exploitation of the poor by the powerful and fraught with war and conflict as Europeans raged over land and religion. It was a time when even Spanish noblemen, whose birthright was membership in the aristocracy, at the summit of that wealthy and expanding empire, suffered the horrors of their bellicose world. Indeed, fighting was in their blood and the empire itself as often as not brought strife to men like Cabeza de Vaca, Dorantes, and Castillo. For they came from a warrior class; men whose badges of honor were their trusty steeds and their weapons, and who, fuelled by pride and prejudice, fought for king and homeland on the frontiers of empire.

In a strange way, Esteban would become an "almost" member of that noble class of valiant, imperial flotsam as he and his companions negotiated their survival in Indian country, far beyond the boundaries of the known world. He too would share the life of a commissioned officer in Spain's imperial venture, in his own way. He too would feel the cut and thrust of bloody butchery in a remote foreign land, fighting for a sovereign he had never seen.

Studies of the Spanish slave trade show that it was usual for a slave to be described in official documents as being from the place where he or she first became the property of a European slave merchant. So, as Esteban was described as a *natural de Azamor*, a native of Azemmour, we can assume that he was sold into captivity there. But the history of Azemmour makes it likely that he was from farther south, beyond the desert, from the Land of the Blacks.

6

"NATURAL DE AZAMOR"

The Slave Trade

THE HISTORY OF Azemmour at the time Esteban was there, sometime between 1500 and 1520, is inseparable from the origins of the Atlantic slave trade. The people of the Mediterranean, Muslim and Christian, European and North African, were fascinated by the almost mythical world that lay beyond the vast, sandy sea of the Sahara Desert. They marveled at the strange stories told by the few merchants who made the long journey across the sands, and they were drawn by the gold and commodities the merchants brought with them. They were especially interested in "black gold," sub-Saharan Africans from the Land of the Blacks who had been sold into slavery and brought for sale in the Mediterranean markets.

For centuries only a few black slaves at a time could be brought north across the desert by the camel trains. But the number rose dramatically when the Christian Spaniards and especially Portuguese defeated the last Islamic kingdoms on the Iberian peninsula. Eight hundred years of the reconquest, eight centuries of continual conflict, had forged a warmongering culture and a warrior aristocratic class. The Portuguese now turned that bellicose and acquisitive approach on Africa, taking control of important Muslim towns such as Azemmour and exploring far down the west coast, raiding for slaves.

For six centuries the Portuguese had ridden side by side with their Spanish cousins, waging holy war, pursuing a crusade against the Muslim presence on the Iberian peninsula. But by the early 1400s, the Spaniards had claimed exclusive rights of conquest over the last Islamic kingdom in Spain, Granada, a place far from the Portuguese frontier, steeped in romantic legend and ruled by indulgent sultans from their opulent palaces of the Alhambra.

The Portuguese aristocracy had a long and noble history of winning their spurs by waging war on the frontiers of Christendom and Islam. But now, without that religious borderland, they were deprived of an infidel adversary and instead went to war with Spain. The fratricidal butchery continued until shedding the blood of fellow Christians began to weigh so heavily on the pious Portuguese monarch and the Spanish king that a truce was called, in 1411. But peace brought a new problem: a generation of Portuguese noblemen who had been bred for war faced the unexciting prospect of living out their lives as gentlemen farmers on their estates at home, idly watching the Atlantic waves break on the rocky shores of their isolated homeland.

A young prince was born into that atmosphere of imminent ennui, almost exactly a century before Columbus returned from America. Prince Henry, "the Navigator," as nineteenth-century historians labeled him, grew into a brave and warlike prince. But he was also a hotheaded fool who promised to give the Portuguese nobility a valiant cause in which they might legitimately invoke God and the Virgin Mary on the battlefield. This quixotic royal maniac was the youngest son of the Portuguese king, John I, by his English queen, Philippa of Lancaster; and in his adolescent lunacy he sired a delusional but noble plan that was conceived in the chivalric spirit of his Plantagenet ancestors.

The mouth of the Mediterranean is guarded by two great limestone crags, which the ancient world knew as the Pillars of Hercules. Exactly seven centuries after the first Islamic army captured the Rock of Gibraltar at the southern tip of Spain, Prince Henry proposed that the Portuguese should capture the fortified port city of Ceuta, the Muslim stronghold on the African shore. It would be a deed that gloried in the pious poetry of its symmetry.

This plan was as ludicrous as it was audacious, but Henry was his father's favorite and somehow managed to convince the aging monarch that such an expedition would be to the greater glory of God. To the blank astonishment of an unwitting Mediterranean world, a Portuguese force of about 18,000 men-at-arms sailed for Ceuta in 1415 and took the town. Henry distinguished himself by charging headlong into battle, oblivious of all danger, an action that cost the life of a trusted royal retainer who was slain saving the young prince. But it was an act that went down in the annals of chivalric folklore. Portugal had a warrior prince more than worthy of the name. The Portuguese conquerors of Ceuta also captured a rich booty, the goods of wealthy merchants who traded across the Sahara Desert.

Ceuta was an important road head for the mysterious trans-Saharan trade routes, which had been become heavily romanticized in Christian minds. Camel trains took fifty days to cross the great deserts. There, beyond the known world, in nearly legendary places like Tombouctou, Muslim merchants engaged in strange exchanges with traders from the Land of the Blacks. There were extraordinary accounts of great quantities of gold bought with piles of salt, that most precious commodity in the heat of the desert. The haggling, it was said, took place in a most alien way. The merchants would find piles of gold at certain places. They would set a pile of salt beside the gold and retire to spend the night. In the morning, if they found the gold and salt still there, they would add more salt, until one morning they would wake to find that the salt was gone and they had been left with the much-treasured, glittering metal. The conquest of Ceuta offered the Portuguese real evidence that those legends and rumors of wealthy lands to the south were true. As a direct result, Henry now encouraged the seaborne exploration of west Africa with much more measured tread. While Henry was at the helm, Portuguese foreign policy steered a steady course of involvement in Africa that directly contributed to the growth of the Atlantic slave trade.

Gomes Eannes de Zurara's famous *Chronicle of the Discovery and Conquest of Guinea* describes those expeditions to west Africa and shows us by example how the Atlantic slave trade began. Zurara explains that when Henry began to organize the first expedition to sail farther south than Europeans had ever explored before, it was difficult to find crews for the ships. Neither experienced nor inexperienced sailors dared to round Cape Bojador (the bulge in the west African coast); they all feared the seas and land beyond it. They had heard rumors of an arid zone described by Aristotle as a place too hot for men, a sandy desert without water, trees, or even scrubby grasses. Some said the seas were so shallow that even a league from the land the water was but a fathom deep. Others said that farther out the waters teemed with monsters that rose, with malign intent, from the darkest depths as though from hell itself. There were fast currents that could carry a ship into the nether regions of the world, from whence it would never return.

But when the Portuguese did finally pass that treacherous cape, they unexpectedly came across a more manageable world. Fifty leagues south of Bojador, they saw footprints of men and camels on the shore. The myth was conquered; they had seen evidence of life and trade beyond fearsome Cape Bojador.

With the phantasmagoric defenses of Bojador breached, Henry launched further expeditions of exploration. In due course, two of his ships—commanded by his young chamberlain, Antam Gonçales; and a young nobleman, Nuno Tristan—came across a group of Africans, a band of fifty or so blacks going about their business near the mouth of a river. The Portuguese attacked them.

Gomes Eannes de Zurara's chronicle reports that the Africans fought fiercely to defend themselves, wielding their assegais, light iron-tipped spears, for, it is explained, "they did not know how to use any other kind of weapon." Indeed, one man fought so bravely he might have been a model chivalric Christian knight worthy of Arthur's table, engaging the well-armed Tristan face to face and defending himself until the Portuguese nobleman overpowered him with his superior weapons.

Of the ten prisoners taken by the intrepid Portuguese during this skirmish, Zurara reports that one was a man "of noble bearing" who offered a ransom of five or six of his own slaves as the price of his freedom. After some bargaining, the slavers exchanged this nobleman and another of their captives for ten more black slaves, some ostrich eggs, a shield of oxhide, and a little gold dust. This was the moment when the Atlantic slave trade truly began, for the young Portuguese adventurers, encouraged, took many more captives from up and down the coast as they sailed for home.

These were in fact the first slaves to be taken on the west African coast and shipped directly to Europe, but it has become traditional for historians to quote Zurara's description of a cargo of slaves brought back by the next slaving expedition as though it were an account of the first black slaves landed in Europe.

Zurara reported that a captain called Lanzarote returned to Portugal with his ships crammed with slaves he had seized on the African coast. He hurriedly sought an audience with Prince Henry and explained that the slaves who had been taken captive "were poorly and out of condition" because of the long time they had spent at sea and "the great sorrow they had in their hearts at seeing themselves in captivity, far from their homelands, without any understanding of what their fate might be." Lanzarote argued that it would be best to bring the slaves ashore as soon as possible. This, he said, should be done in the early morning before the heat of the day. He suggested that they should be taken to a field just outside the city gate and then be divided up into five groups.

The following day, August 8, 1444, very early in the morning, the seamen made ready their boats and slowly began to ferry the captives ashore. Zurara described the scene as astonishing. Among the slaves were some who were white, "indeed, fair to look upon, and well proportioned." But others were "less than white, like mulattoes. Others still were as black as Ethiopians," he wrote, describing them as "so ugly, both in features and in body, as almost to appear (to those who saw them) the images of a lower hemisphere."

"But what heart could be so hard," he continued, "as not to be pierced with piteous feeling to see that company? For some kept their heads low and their faces bathed in tears, looking one upon another; others stood groaning with great anguish, looking up to the heavenly firmament and fixing their gaze upon the skies, crying out loudly, as if asking help of the Father of Nature. Others struck their faces with the palms of their hands, throwing themselves at full length upon the ground; others made their lamentations in the manner of a dirge, after the custom of their country. And though we could not understand the words of their language, the sound of it right well accorded with the measure of their misery."

Their suffering was to become still worse. The men in charge of dividing up the captives arrived and began to separate one slave from another, "in order to make an equal partition of the fifths." Making no attempt to keep families together, they parted fathers and sons, took husbands from wives, and sundered brothers from brothers. No respect was shown for the ties of the past and "each fell where his lot took him."

"O powerful fortune," Zurara lamented, "that with thy wheels doest and undoest, compassing the matters of this world as pleaseth thee, do thou at least put before the eyes of that miserable race some understanding of matters to come; that they may receive some consolation in the midst of their great sorrow. And you who are so busy in making that division of the captives, look with pity upon so much misery; and see how they cling one to the other, so that you can hardly separate them.

"And as often as they had put one group of captives in one place, the sons, seeing their fathers in another, arose with great energy and rushed over to them. Meanwhile, the mothers clasped their children in their arms, throwing themselves flat on the ground to cover them, where they took the beating of their tormentors with little pity for their own flesh so long as their children were not torn from that embrace."

The whole population of the town abandoned work for the day as word spread of the strange sight just outside the city walls. "The field was quite full of people, both from the town and from the surrounding villages and districts," many of whom were moved by what they saw, and while some were weeping and others were dividing up the captives, others "caused such a tumult as greatly to confuse those who directed the partition."

These scenes eloquently convey the terrible beginnings from which the Portuguese built the Atlantic slave trade. The high value of that human cargo intensified the greed of European merchants, who now eagerly prepared for future plunder. As the trade expanded rapidly, the Portuguese established their now notorious "factories," fortified and garrisoned trading posts built at strategic places on the west African coast and on the islands in the Bight of Benin. These were the essential infrastructure of the trade, warehouses in which to store human beings. Feeding the men of the garrisons and the thousands of slaves imprisoned in the "factories" was a major problem for which Azemmour provided a ready solution.

Azemmour was an excellent port and the largest town in the Moroccan region of Dukkala, an agricultural land with abundant wheat fields and market gardens—the breadbasket of the western Maghreb. Exports from Azemmour and other Moroccan ports, would, quite literally, feed the slave trade, providing the "factories" with enough to eat. It is no surprise, then, that Esteban should have ended up in Azemmour.

TODAY, AZEMMOUR IS a small, tranquil, atmospheric town on Morocco's Atlantic coast, a place of infinite afternoons and lazy backgammon evenings spent in narrow streets and shady courtyards. It is announced by a small and dusty signpost that lurches precariously into a dry ditch beyond the roadside embankment, an old world place with all the charm of authenticity. The river flows idly by the crumbling city walls that Esteban once knew; this is a place untouched by the tourist razzmatazz of nearby Marrakech or neighboring Essaouira. But this modern sleepiness belies a distant past that was both vibrant and violent. Azemmour was once a frontier trading post that uncomfortably straddled the great religious rupture of the pre-Reformation age, the rift between Christendom and *Dar-al-Islam*, the Muslim lands. It was a place of mystery, a gateway to a world of wonders, one of the African terminuses for the great sub-Saharan trading routes that brought gold and slaves to the threshold of Christendom.

We will never know how Esteban came to Azemmour. In all likelihood he was brought as a slave, either in one of the trans-Saharan camel trains of the Muslim merchants or, more likely, aboard a Portuguese slaving ship plying the coastal trading route. Perhaps his parents had been taken as slaves and brought to work in North Africa as household servants, laborers, or craftsmen and Esteban was in fact born in in the town.

By contrast with those obscure beginnings, the story of how Esteban came to Spain from Morocco is more easily surmised. In the late 1510s, the great breadbasket of Dukkala, with Azemmour at its heart, was struck by famine and pestilence. Esteban almost certainly arrived in southern Spain as a refugee from these terrible natural disasters that destroyed the region forever.

Drought came first, in 1517, and then the resulting failed harvests brought the terrible suffering of famine to the region. Wheat and hardtack biscuit were shipped to Azemmour from Spain, but no sooner had recovery begun than drought struck again, bringing more failed harvests and further famine. Now, an incessant, distressing drama was played out against the backdrop of the threat of plague, which was then spreading steadily across northern Morocco, perhaps arriving on the very same boats that had first brought food and relief from southern Spain.

A chronicler of the period tells of how, in 1522, the first Portuguese merchants arrived at Azemmour, as they did every year, to load their ships with the abundant fish that the inhabitants took from the river and salted for this trade. But these merchants and sea captains soon changed their plans. The desperate people of the town crowded around their ships, pleading that the merchants take them as slaves, willingly selling themselves into captivity for the price of a meal. The captains filled their ships and set sail for Lisbon and Seville carrying their unexpected human cargo.

Up the coast, in Asilah, a wily merchant saw this as an opportunity to help out a needy widow who had no doubt caught his eye. He had just negotiated a deal with some well-connected businessmen and was about to set out for Azemmour to make the most of the opportunities offered by the misery of the people there. As a favor to the widow, he offered a place on his ship to her youthful son, Bernardo Rodrigues.

Fortunately for us, the young Bernardo was so struck by what he saw that he wrote a detailed description of his experiences. He was shocked. Fathers sold their sons and daughters, and brother sold brother. Bernardo said this was so contrary to human nature, so perfidious and unheard of, something so etched on his memory, that he knew he had to record what he had seen.

At Azemmour, Bernardo saw the massed starving people for the first time. Emaciated families were dying like beasts alongside the camels and cats that others had slaughtered in the vain hope of a meal. But soon, many ships arrived, bringing reports that the situation was even worse at Safi and Azemmour. And as evidence of this, the ships had come packed with people from those towns who had sold themselves to avoid starvation.

Bernardo's captain set sail immediately for Azemmour. As soon as they arrived they saw the river packed with ships that seemed to overflow with beautiful girls, for Moorish women were much prized as concubines by the Portuguese. Bernardo boasted that when he was at Azemmour, for a few nickels he had bought a tall Moorish girl who was less than twenty-five years old. She was very pale and beautiful, much to his taste, and he had kept her at home ever since. No doubt she had sold herself to this conceited young man for the sake of her six-year-old son.

In hope of adding to his earlier purchase of a mistress, Bernardo now set out for the suburbs of Azemmour. There, among the dried-out irrigation systems and parched fields, he found a man who was selling his daughter and his granddaughter, "both good looking." Bernardo paid thirty-two nickels for the one and twenty-eight for the other. With the transaction over, the desperate man nobly offered the new master of his womenfolk a little roast camel meat, perhaps hoping to soften Bernardo's heart. But Bernardo did not eat the flesh, perhaps out of pity for the hungry man, perhaps out of prudence toward his own stomach.

Bernardo made one last purchase, a young man of noble disposition, for whom he paid sixteen nickels. These sales, he explained, "should be enough for you to understand the terrible hunger."

It will almost certainly remain impossible to prove that Esteban was among the desperate refugees who escaped famine in Azemmour on ships manned by bargain hunters like Bernardo. But until a better and more probable explanation is argued from the evidence available, this must surely be considered his most likely route to Spain.

7

JEREZ

1522

As Esteban embarked on a ship overcrowded with refugees at Azemmour in 1522, the scenes around him were similar to television images familiar in our own times: hungry, nearly broken Africans gathered together in search of survival, their faces ghostly with bewilderment as they abandon themselves to an almost lifeless hope. Once on board, Esteban and his companions resigned themselves to the false promise of a new dawn.

For those lucky enough to have been earmarked as ideal commodities for sale in the slave markets of Lisbon or Seville, there was perhaps some reason to be optimistic about saving their lives, although at a considerable cost. They, at least, might be bought by masters who were required by law to ensure that their slaves were properly clothed, fed, and sheltered. Many of those who embarked may even have had some idea of their legal rights. Whether many masters had either the inclination or the wherewithal to comply with those laws in such a woeful time is another matter, but some certainly did. For a few of the refugees, then, their hope was not in vain.

But for many more, the fabled Spanish shores of which they had heard so much talk from the merchants and mariners of Azemmour were merely the tragic epilogue to a horrible story. Southern Spain is only eighteen miles from Ceuta and Morocco, and pestilence and dearth had struck the Spanish too. Again, we must rely on conjecture to establish Esteban's likely story, but a young Spanish chronicler, Juan Daza, describes the terrible destiny of the refugees from Azemmour. In reality, their prospects were hardly better in Spain than they had been in North Africa.

Juan Daza, like Cabeza de Vaca, was from the small city of Jerez de la Frontera, famous for its sherry wine and Thoroughbred horses. The two men were close in age and must have come across each other from time to time in the claustrophobic village atmosphere of Jerez. The apocalyptic horror and catastrophe that engulfed the place at the time Esteban arrived in Spain is described by Daza in a naive writing style that betrays his youth but is confident and authoritative in tone.

Daza recorded that in 1522 there was not enough grain to sow for a harvest the following autumn. The charitable hospitals were filled to overflowing with the sick and the starving. People resorted to all manner of foodstuffs, and the streets of the city rang with cries of mendicant monks, begging for the poor with their plea: "For the love of God!"

With them went a great crowd of youngsters, calling out the heart-rending lament of the starving poor, "I am dying of hunger!" At night, these miserable creatures slept in the streets, the haunting moans of their private anguish punctuating the slumber of more fortunate citizens. In the morning there were many dead, but there was no one to bury them.

Daza tells us that there was no one to shed a tear for these beggars and no one would give them food; nor was anyone prepared to stop them from stealing it, even though it was scarce and expensive. They stole figs, walnuts, and chestnuts and took bread from the bakers and cheese from the cheesewrights, and even the sheriffs and their deputies turned a blind eye for there was neither dog nor cat nor donkey left alive in the stricken town—the hungry had eaten them all along with other things that were far worse. In the countryside men and women lived like beasts, eating thistles, weeds, and other poisonous herbs and grasses that killed them.

Then, in the cycle of disasters, plague struck Spain again in the late spring of 1522, and the bodies piled up in the streets of Jerez and other Andalusian cities. The churchyards were full and no one could enter the churches because of the stench of rotting corpses.

"It was shocking to see the bodies of the Moorish slaves, men and women, piled up in the city rubbish dumps where they had been buried out of necessity . . . and awful to see how many of these Moorish carcasses there were," Daza explained. "And the reason that there were so many Moors here is that last year there was great famine over there, beyond the sea, in the regions of Safi and Azemmour. The Moors themselves approached the Christians and pleaded with them to take them captive and to bring them to Spain."

So many were brought in 1522 that their value fell to practically nothing, not least because of their terrible health.

"More than six thousand of these souls were brought to Spain, but it was awful to see them arrive because they were so weak and misshapen from the great hunger they had suffered," Daza reported. "Nearly all of these aforementioned Moors perished, and very few escaped the plague, the fiercest pestilence ever witnessed by man, which took hold in May and lasted until the Day of Saint John the Baptist (June 24)."

It seems likely that Esteban was one of the few sub-Saharan Africans who managed to escape Azemmour alongside these Moors. But that may have saved his life, because, as we have seen, Spaniards tended to prefer *negros* to *moros* as slaves. Esteban would have stood out as a more valuable commodity than the starving Moorish refugees around him.

These were desperate times, marked by awful fears and ghastly deeds. Juan Daza describes how, in 1522, at the orders of the chief sheriff of Jerez, an emaciated, half-starved young lad was paraded naked through the streets, riding on the back of a donkey. From time to time, as the small procession stopped here and there, the boy was tortured with red hot tongs while his grisly crime was described by the town crier. These agonies were abruptly and brutally brought to an end when he was "quartered." Daza is not specific, but if the punishment was meted out in the usual fashion, then the criminal—Daza judged him to be "no more than eighteen"—was disemboweled and his genitals were cut off while he was still alive. He was then beheaded, and finally his cadaver was cut into four parts. Daza tells us that the four parts were then gibbeted on four roads leading into the city, while his head was posted on a fifth. Daza watched it all, along with a fascinated crowd. No doubt if Esteban was in town and had the opportunity to see it, he would not have missed such an exciting public entertainment.

The young victim of this vicious justice had, however, committed a heinous crime and had violated an ancient taboo. It was April and the warm days of springtime were already giving way to summer in that parched year of failed crops, when the boy, almost crippled with hunger, approached a humble cowshed. He was in luck, for one of the local cheesewrights was preparing to spend the night there.

"For the love of God," begged the boy, "may I shelter here for the night and get some sleep?"

The compassionate cheesewright assented, and the boy pleaded for food: "Give me a mouthful of bread, for it is eight days since I had a bite to eat and I'm starving to death."

"Congratulations! You are in luck," said the good cheesewright, and he gave the boy some bread and milk, which was what he had in the cowshed. As the boy ate, the benevolent man made up an extra bed for him.

In the morning, as soon as the two companions had awoken, the cheesewright gave the lad a breakfast of bread and milk.

"Stay as long as you like in this cowshed," said the host, for he was overcome with emotion at the sight of the such a weak and perilously thin youngster.

As the two of them chatted, the cheesewright became drowsy again and slowly dropped off to sleep.

The boy suddenly seized a hoe and smashed the good man over the head. With a knife, he quickly ripped out the man's intestines. Then he butchered his benefactor, slicing off those parts he most fancied for eating, and filled some saddlebags with the man's flesh.

He spied a mare nearby, brought her to the hut, saddled her, and buckled on the saddlebags. But as the murderer was ready to make good his escape, the chief cattleman appeared. The boy panicked. Petrified, he failed to mount the horse. The cattleman saw the blood.

"What's that?" he asked.

In his agitation, the boy was unable to hide the facts of his crime. "I killed the cheesewright," he confessed. "Then, I butchered his body."

It seems from Daza's account, which he perhaps heard directly from the cattleman, that both of them were shocked and aghast at what had happened.

"I did it for the meat," the boy explained. "For the meat and for no other reason."

The cattleman grabbed him and quickly tied his hands. He brought his prisoner to the city and he also brought the grim evidence in the saddlebags. The prisoner was thrown into jail. The sheriff was sent for. The prisoner was sentenced to death.

And, when a few curious citizens of Jerez, with Daza no doubt among them, asked this eighteen-year-old why he had committed such a horrible crime, he replied: "The devil said, Get up and kill him!" But he confessed that it was "also so that I could eat my fill of the cheesewright's flesh, for I was starving to death."

This was the world into which Esteban stepped as he first arrived in Spain from famished Morocco.

CANNIBALISM IS AN almost uniquely emotive subject. It has left a deep scar on the human mind and it lies at the heart of western European mythology, for it is central to Christian worship. Catholics believe that the bread and wine of Holy Communion truly become the flesh and blood of Christ at the moment of consecration. This mysterious miracle of transubstantiation is linked to the idea of Christ as a sacrificial lamb, an idea that is rooted in the more prosaic tradition of sacrificing a lamb for Passover that was so important to Christ's own religion, Judaism. And, what is more, it is widely held that the tradition of slaughtering an animal in homage to the gods was little more than a civilized development of an original tradition of human sacrifice.

The Spanish conquistadors found such human sacrifice in Aztec Mexico, a barbaric ritual which they claimed went hand-in-glove with cannibalism, although this has been eloquently disputed. But Europeans held deeply rooted beliefs about cannibalism among "savage" Americans. In fact, the word "cannibal" is a corruption of "Carib," from which we also derive "Caribbean." This belief was a direct result of Christopher Columbus's deluded perception of the New World, for he quickly became convinced that the aggressive peoples he came across on his first voyage in 1492 deliberately waged war against their pacific neighbors in order to eat them. But although Columbus may have sown the seed of this European mythology of Carib man-eating, others soon enthusiastically embellished the idea. The inhabitants of the New World were soon simplistically characterized and classified as either bad or good, depending on whether Europeans thought they were the cannibals or the ones who were eaten.

The Spanish monarchs responded. They signed a decree that enshrined this fantastical distinction between good and bad in law by establishing that it was legitimate to enslave cannibals, whereas the good Indians who did not eat human flesh were to be kindly treated. This was a well-intentioned act of pious rulers who had no understanding of the real situation. But, by signing that decree, they gave every Spanish conquistador good reason to allege and even to believe that the people he persecuted were man-eaters.

How many Americans in 1492—or how many humans throughout history, for that matter—were habitual man-eaters, if any, has been the source of considerable debate, often undignified, among scholars, but it is a question of only passing interest here. It is of paramount importance, however, that Europeans considered cannibalism so atrocious that it defined the difference between peoples who could be considered human and those who should be treated as beasts, the difference between those who should be treated as royal subjects and those who should be enslaved.

Therefore, cannibalism by a European was not merely an especially repellent and revolting act that violated all sense of what it meant to be civilized. It was much worse because it did violence to the essence of what it meant to be human. And so the eighteen-year-old in Jerez had been on the verge of literally bestial behavior, of committing an act of animal atrocity. He had betrayed his humanity.

In fact, the intended crime may have been worse still. The year before, an epidemic had struck the cattle of the region, leaving the animals weak and vulnerable to rustlers and thieves. Many people were driven to mild, acceptable dishonesty by their hunger, killing a beast in the field and taking the meat home for the family pot. But there were many others who stole out of greed, Daza reported, selling the meat for a profit. These malefactors, having learned to ply their delinquent trade by stealing cattle the previous year, now turned their attention to the donkeys and horses that had been put out to pasture. Could it be that the eighteen-year-old victim of sixteenth-century justice intended to sell the butchered cheesewright?

By the middle of July the plague had abated. It is possible that Esteban remained in Jerez, where the first religious brotherhood to be founded in the city by black Africans would shortly be established, in 1527, the year Narváez sailed for Florida. But it seems more likely that Esteban now headed north, to Seville, which was Europe's gateway to America and the commercial powerhouse of the Spanish empire.

8

SEVILLE

1522–1527

THE GREAT PORT city of Seville, in the deep south of Spain, was at the heart of all European trade and traffic to America, the port on which America then depended almost completely for contact with the Old World. As Esteban's journey brings him ever closer to America and the Spanish colonial world, the probable facts of his life become clearer and more certain. We can be sure, for example, that he passed through Seville, because almost all passengers who sailed for the New World were legally required to register themselves and their slaves at the offices of the "House of Trade," which were in the city. Moreover, the Seville market was where hopeful conquistadors like Andrés Dorantes bought the slaves they took to America, personal servants purchased to play the role of squire to the master's knight-errant.

When Esteban entered the city through one of the many gateways that pierced the massive ramparts, he would have found himself hurled headlong into a human boiling pot. This was a quickly growing city, soon to become the richest and one of the most populous places on earth. The good and the bad, the beautiful and the ugly, all enthusiastically came to Seville in search of something—usually money and often adventure. It was like some Vulcan's forge where the ore of modern European life was smelted, the threshold of the New World, a social and cultural cauldron, the devil's kitchen. And Esteban was thrown into this rich recipe as the very humblest of ingredients: a newly arrived slave.

By contrast, today, in Seville, History wraps herself around you in an intimate but deceitfully sweet caress. Nowadays, one comes across some unpredictable piece of the past at almost every one of the many turns which punctuate the tortuous lanes of old Seville. As you stroll through the shady streets to the north of the cathedral and the old Jewish quarter, escaping the noonday heat, you come across four marble columns that once formed part of a Roman temple and which stand abandoned in a vacant, flooded lot. Not far away, the elegant arcading of the former Arab bathhouse now forms the entrance hall of a popular pizzeria. Beyond that, a small church with perfect proportions and rococo aisles was once a synagogue, but well over 500 years ago. The fashionable nightclub and flamenco bar that is so difficult to find in the labyrinthine streets behind this church was once a warehouse where charcoal burners smoldered their wares. And the noisy tapas bar next to a disused theater occupies a ground-floor room of the former royal mint, where the gold and silver brought from Mexico and South America were made into bullion and coins.

The intimacy of these surprises, these almost archaeological encounters with the detritus of history, takes place against a backdrop of the grand buildings erected by church, state, and commerce. The bulk of the city hall, the *Ayuntamiento,* still exudes power; the merchants' exchange, the *Casa Lonja,* quietly dominates the main entrance to the town. Here, as almost nowhere else in Europe, the church towers are still among the tallest buildings. The highest point of all is the tip of the cathedral tower, the Giralda, once the minaret of the main mosque, where the muezzin sang out his call to prayer. Now, the baroque bell tower performs the same function for the Catholic faithful.

These vast monuments signal the now long-lost affluence of empire. And, in a city unencumbered by modern skyscrapers, which cast their shadows upon ancient neighbors, these monuments remain the impressive sentinels of an imposing past which continues to preside over the parochial present.

If you enter Seville from the north, once you have driven through the ranks of concrete apartment blocks built beyond the old city walls, you soon come across the imposing Hospital of the Five Wounds, built in 1600. There is a strong sense of the presence of history in this austere stone building. Today it houses the Parliament of Andalusia, but it was built to shelter the poor and for the care of the sick. Despite its great beauty, the place is testimony to the fact that an enormous building was required to tend the terrible privations of many broken souls. It was one of two giant hospitals that gave solace to the needy, along with many smaller institutions that did the same. Wealth and poverty were uneasy bedfellows in sixteenth-century Seville.

On the far side of the traffic-laden boulevard of the ring road, opposite the Hospital of the Five Wounds, the peripheral modern city ends. It is as though the serried ranks of apartment blocks have come up against the grand swath of the crenellated curtain walls with almost all their medieval towers and turrets intact. These strong defenses that still stand after 1,300 years have stayed the hand of progress. Now they are monuments, but they were built by the Moors to defend their city from attack, first by the ruthless Vikings, then by the Spanish crusaders who finally conquered them. The medieval city was born of war and splendor.

Spain's golden age of empire has in some ways stood the test of time in the potent sense of bygone glory that now dominates the landscape of this peaceful, pious, provincial capital. The Virgin Mary is still worshipped as fervently as she was in Esteban's time. As in Esteban's time, there is deep devotion to the dogma that Mary herself was immaculately conceived, thanks to the miraculous "kiss" between her mother, Saint Ann, and her father, Saint Joachim. Easter and Holy

Week bring crowds into the streets to witness the religious processions that wind their way through the city by day and by night. The dirt of centuries clings to the ceilings of wineshops, bars, and taverns on the streets leading to the river. The city is still dominated by landowning families, farmers on a grand scale, the originators of agribusiness. It remains a place of bullfights, flamenco music, and Gypsy bacchanals.

IN SEVILLE, IN the spring of 1522, the despairing city fathers appealed to the royal authorities for help with the devastating famine and plague. They reported that the destitute and the paupers of the surrounding countryside had descended on Seville in such numbers that the hospitals were suddenly too full to accommodate the sick and the hungry. Esteban was almost certainly among them. The good burghers claimed to have done what they could, releasing as much food as possible from the official reserves, but even so, more than 500 emaciated cadavers had already been found in the streets and the death toll was rising steeply. It was a heartfelt plea for assistance from local oligarchs appalled by the suffering on their doorsteps and terrified that this miserable mass of unhappy humanity would soon bring disease and death to their city.

Whether Esteban arrived in Seville at this time by his own wretched volition or was brought to the city already enslaved, the simple fact of being a *negro* was enough to mark him as a slave. Without documentary proof of his freedom, a black man in early-sixteenth-century Spain was assumed to be chattel. By contrast, *esclavos blancos*—the Moors, Berbers, and Turks—were usually branded on the face, sometimes with the name and profession of their owner, but more often marked with an S and the symbol of a nail, †, effectively a sort of dollar sign. These marks signified the Spanish word *esclavo*: S for *es* and the nail for *clavo*.

Within days, if not hours, of his arrival in the city, Esteban found himself in the central market, one of the most unusual bazaars in history. There he and hundreds of other woeful figures were in due course put up for sale alongside a seemingly endless variety of other merchandise proffered by the city traders to passersby.

The merchants and contractors of sixteenth-century Seville conducted their business in and around the cathedral. They gathered on a continuous flight of steep stone steps called the *gradas*, which today still encircle the massive Gothic church, as though trying to raise something sacred above the profane street life. Ironically, these iconic steps provided those businessmen with a sort of sanctuary from the profanity of local taxes, because technically, the *gradas* were within the precincts of the cathedral and so came under the jurisdiction of the church, and not the city hall.

An Italian ambassador who traveled through Seville in the 1520s described the wide marble pavement above these *gradas*, fenced off by a series of columns connected by sturdy iron chains. There, merchants and other gentleman promenaded, while all about them the market traders sold their wares, spilling over into the nearby square, which was constantly thronged with crowds. There, the Italian reported, many were tricked by merchants or conned by grafters and cardsharps.

A contemporary Sevillian commentator explained that traders sold their goods by auction there on each and every day of the year, with the exception of Sundays and religious festivals. "As many riches as you could imagine" had their price, but he was particularly struck by the quantities of sumptuous clothing, often embroidered or otherwise decorated with precious metals. He also noted that arms, armor, and slaves were auctioned there as well. Nearby, "in front of the *gradas* there were various expensive silversmiths' shops on the right-hand side; and beyond, the buildings of the Bank of Seville, where such an infinite amount of money is exchanged that it is the most liquid bank of any of which I have heard."

It was from these *gradas* that Pánfilo Narváez or his captains and agents proclaimed his grandiose adventure. On these steps, hopeful predictions and false promises were made, hands were shaken, contracts were agreed and signed. Esteban was either displayed for sale standing on the top step or paraded through the crowded street below.

The barely controlled mayhem—the hectic, almost hedonistic atmosphere of this palpitating commercial heart of Seville—led many of Esteban's contemporaries to describe the city as a Babylon, a town of taverns that attracted prostitutes and other sinners. It was certainly a Tower of Babel, a place where hundreds of nationalities spoke in myriad different tongues, all united by the common language of commerce, a language that at the *gradas* rivaled the power of religion.

At Azemmour Esteban had become used the sight, sound, and smell of the many Italians, Portuguese, and Spaniards who came and went among the crowds of Moors and Africans. But as Esteban was led, shackled or bound, through the streets to the cathedral he must have been immediately struck by the variety of exotic complexions, by the otherness of the white European world. He saw a sea of foreign faces, some burned by the brutal sun, others as fair as the faces of fairytale princesses, with golden hair and pink cheeks. Among the Spaniards there were Germans, Flemings, and Frenchmen, traders and hopefuls from all across Europe. No doubt even the stench of a unwashed mass of Europeans seemed a rank novelty when compared with the human odors to which he was accustomed.

He may even have gained his first glimpse of an "Indian," a native of the New World. The Italian ambassador remembered how, at Seville in the spring of 1526, he had seen some youths from America, the sons of important people. They had come to learn about Spain, he reported, accompanied by a friar who had been preaching in the New World. He described them as being bright and lively, with jet-black hair, long faces, and Roman noses. They dressed in their own style, wearing a kind of doublet or smock. This observer was most impressed by a game of ball that they played. The ball was made from a knot of wood the size of a peach, or perhaps bigger, which was very light and bouncy. Most amazing of all was that according to the rules of the sport they could use neither their hands nor their feet. Instead, they kept the ball in the air and passed it from one to another using only their bodies, and they did this so skillfully and at such great speed that the Italian was quite astonished. At times, he explained, they even lay down completely to return a pass.

But as surely as the human eye is drawn to the familiar, so Esteban's attention soon turned to the many African faces in the crowds of Europeans. There were Wolofs, Mandingas, Guineans, and many more from all over west Africa, including many whose characteristics were almost as foreign to Esteban as those of the Germans and the Flemings.

Seville had a long tradition of black slavery stretching back more than three quarters of a millennium, to the time of the Moorish sultans. In Esteban's time so many Africans lived in Seville that foreigners compared the city to a game of chess: because it seemed to be populated by as many blacks as whites.

Information from later censuses suggests that as many as 1,500 black African slaves may have been resident in Seville when Esteban arrived. There were also many *horros* and their children: freedmen, *horros* were former slaves, often granted their freedom in the wills of their Spanish masters. And there were also countless slaves destined for the Indies who passed through Seville, where their export was licensed and the relevant taxes were paid at the House of Trade. It is impossible to know how many black Africans lived in Seville in the early 1520s, but there were many: perhaps as many as 3,000, almost a tenth of the already considerable population that was on the verge of an explosive increase which would make Seville one the largest cities in Europe by the end of the century.

What is more, this black minority was highly visible: many blacks worked as stevedores in the port, or as porters, as street vendors, or even as constables, a generally unpopular job.

And, while most slaves worked as domestic servants, they were not usually confined indoors, but were instead often out on the street, working as gofers, accompanying their masters, running errands, and the like.

The life of enslaved servants varied from household to household. The greatest of all European writers, Miguel de Cervantes, describes some who slept in the stable, along with the master's other living chattels, but there were others who benefited because they were treated as luxury goods. Rich Spaniards dressed their household slaves in outlandish finery as symbols of their wealth.

One flabbergasted Flemish scholar, newly appointed as a tutor at the Portuguese court, described how noblemen went into the streets in the town of Évora. Two slaves marched in front of the master, who was on horseback. A third was there to carry his hat, and a fourth to carry his cloak, lest it rain. A fifth held the horse's bridle, a sixth the gentleman's silk slippers, a seventh his clothes brush, and an eighth a cloth to rub down the horse while the master attended Mass or chatted with a friend; and yet another, the ninth, brought up the rear, taking good care of his master's comb.

The wealthy of Seville were equally ostentatious and similar displays of arrogant opulence were a familiar sight on streets as the city's bourgeois nobility waged a permanent war of display among themselves. According to one theory, Europeans may have begun to make mannequins of their African slaves and bedeck them with jewels and luxurious clothes because such gaudiness was traditional in many of the societies from which the slaves had originally been taken.

While the Flemish scholar may have been shocked at the sight of a train of richly dressed black servants, it seems unlikely that such things would have seemed unusual to Esteban. From time to time in

Africa he must have seen overattended masters, lords, and princes, white, black, and Moorish.

We know from Juan Daza, the young man who described the plague and famine at Jerez, that the price of slaves had fallen precipitously by the early 1520s because of the many refugees from Azemmour and Safi. But we also know that, in general, Spaniards seem to have preferred *negros* to the "white" Moors and Berbers, who were more troublesome. So Esteban must have stood out as especially desirable in a market awash with Muslim infidels. He had also probably learned Portuguese at Azemmour, and perhaps some Spanish; and he may have worked as a domestic servant or in some trade. With the Portuguese presence at Azemmour, he may have already been a Christian. With these advantages, he was no doubt quickly sold at a good price.

The burgeoning African population of Seville presented the authorities with a novel problem. Individual black men were a familiar sight, because there had been black slaves and servants in the city for centuries, but en masse they did not fit any of the existing ethnic categories by which Spaniards understood the cultural diversity of their world. They were certainly not Spanish, but neither were they Moorish, Muslim, or Jewish. In fact, they were almost all Christians because African slaves were immediately baptized on arrival in Spain or Portugal.

There was a range of attitudes toward Africans in Spain, and many Spaniards evidently saw them as fellow human beings. Marriages between whites and free blacks who had become assimilated into Spanish society were usual in the lower classes. One very reliable modern scholar has suggested that Spaniards had a reputation for being unusually kind to their slaves, and she may be right. However, the notion appears to based on the passing remarks of an early modern French commentator, taken somewhat out of context. No doubt many Spaniards did in fact treat their slaves well, sometimes out of humanity, sometimes for much the same practical reasons that may have caused them to cared for the rest of their property. But there were many others, including men of considerable influence, whose opinion of black Africans could lead only to abuse.

Of these, Bartolomé de Las Casas is perhaps the best-known. Because of his humanitarian complaints about the treatment of Indians by the Spanish conquistadors, which helped to galvanize the ideas of men like Zumárraga, Las Casas has rightly gone down in history as the "defender of the Indians." But in his zeal to save the Indians, he eloquently argued the case for sending African slaves to the New World instead.

It has become commonplace to forgive him this aberration because he repented as soon as he saw the evil effects of slavery on black Africans. And there is no question that Las Casas was basically a good man, compared with his peers, even from the perspective of our own age. Which is why it is absolutely essential that history remember he was also an influential advocate of the transatlantic slave trade at its inception. He thereby contributed, right at the start, to the centuries of evil to come. Only if his reputation is made to bear this burden can we understand that it was commonplace in his time for even the most humane and compassionate Europeans to think of black Africans, at least in the abstract, as in some way subhuman.

The argument for enslaving Africans was simply put by another politically influential intellectual, Peter Martyr, chronicler to the court of the Catholic kings, Ferdinand and Isabella, the man who coined the term "New World," *orbe novo*. Peter Martyr argued that for all that Natural and Canon Law might decree that all human beings should be free, Roman Law defined an important exception, and in practice enslavement had become established as acceptable. "Long experience has effectively shown the need for the enslavement of those who by their nature have a propensity to abominable sin and who would return to their evil way if not guided and tutored."

Another influential Spanish writer, Tomás de Mercado, took this argument one stage farther and used the very process of trafficking in human beings as a justification for slavery. Because African blacks are "savage sinners," he wrote, "they commit ghastly crimes.... They act not out of reason, but in accordance with their passions.... It should surprise nobody to learn that they treat one another abominably, selling each other, for they are uncivilized, wild, and savage ... and treat each other like animals." He then explains that they not only sell their children when in dire straits but also needlessly, "out of irritation or anger at some trivial insult or bad behavior."

The view of some, then, was that as a black African Esteban was "uncivilized, wild, and savage." Black slaves certainly had a reputation for being thieves, drunkards, and libidinous. Given such attitudes, it is hardly surprising that in Seville during the sixteenth century, the growing and very visible black population was cause for paranoia among the city fathers. In response, they produced various municipal ordinances intended to restrict the activities of slaves. Slaves were not allowed to carry arms except in the company of their masters, and there were limits on the number of slaves allowed to assemble in public. The municipal oligarchs in Seville were especially concerned about slaves and alcohol. In 1569 they issued an apparently draconian bylaw that excluded slaves from all taverns and public eating houses unless specifically given permission by their masters:

> Because the many blacks and Moors who are slaves in this city have plenty of opportunity to go to taverns and public eating houses to eat and drink they become drunk and disorderly, behaving arrogantly. When drunk they commit crimes, for which their owners have to pay.... What is more, the slaves are so addicted to the taverns and the eating houses that they steal money or clothes from their owners, even taking blankets and harnessing for the horses and mules, or anything else lying around.... And the tavern owners, their wives, daughters, and others who go to the taverns fence these stolen goods for them.

In reality, this was simply one aspect of a widespread distrust of taverns, inns, and public eating houses—those much frowned-on but essential services in a city filled with sailors, merchants, and transients. There were continual attempts to control such hostelries, none of which seem to have been very successful. There was even a law forbidding the sale of both food and wine in the same place, on the grounds that a hungry husband was likely to stop boozing, leave the tavern, and return home for dinner. A curious consequence was that itinerant food sellers and cooks went from tavern to tavern offering their wares. The drinkers would simply call these hawkers in and have them prepare food.

But one Italian visitor to the great Portuguese seaport of Lisbon seems to have approved of some of the more rowdy Africans' behavior, commenting that "while the Portuguese are always sad and melancholic . . . the slaves are always happy, they do nothing but laugh, sing, dance and get drunk publicly in all the town squares." No doubt many in Seville felt the same way as this Italian.

The paranoia experienced by the fearful urban elite led to the partial exclusion of African slaves from the society of which they were a vital and integral part. The result was a distinct "slave" society parallel to the white world. And as Esteban began to find his feet in the metropolis, he must have quickly realized that among the whites, he was simply a *negro*. The European hierarchy decreed that his proper place was among the African slaves. Within that world of black slaves, however, Esteban could be more than just an "African." He could shed that European label and establish his own ethnic and cultural identity.

Esteban no doubt soon found his way to the small plaza near the church of Santa María la Blanca, where the black population of the city tended to gather, much as the Africans of Lisbon today congregate near the church of San Bernardo. In more tolerant Muslim times, Santa María la Blanca had been a synagogue. It still stands where it was then, just inside the walls, by the Gate of Meat, so called because it was the entrance closest to the municipal slaughterhouse. This gate was also close to the outlying suburb of San Bernardo, a desperately poor working-class district, surrounded by swampy land, where many black Africans lived.

The *negros* who gathered near Santa María la Blanca were satirized in a literary interlude, a humorous vignette, once mistakenly attributed to Miguel de Cervantes. Writing some eighty or so years after Esteban's time, the anonymous author rather unoriginally described Seville as being like Babylon, a place where there was a story to be told at every hour on every street corner.

A character in this comedy recalled a story told by a student who claimed "to have been passing through the parish of Santa María la Blanca and that small square where endless blacks gather together." The student sidled up to a group of *negros* who were deep in conversation, and began to eavesdrop. They were, he discovered, extremely courteous and formal with each other. Then one of them asked another, "in his half-baked Spanish," "Your grace, tell me, is it true that your master has sold you?"

"Yes, sir, he has indeed," replied the other.

"And how much did he sell you for?"

"He sold me for one hundred twenty ducats."

"Good lord, how expensive! Your Grace is not worth that much, for you cannot be worth a penny more than eighty ducats!"

The other, far from getting angry, then explained that he had married a black girl from the area against his master's wishes. His master, angry at such insubordination, told him that he could only sleep with his wife on Saturdays, to which the slave had responded by asking how many Saturdays there might be in the week.

When the master told him that there was only one, the slave had replied that he would be happier with three Saturdays in the week. Moreover, if he was only allowed one Saturday he would have a word with the authorities and request that justice be done. His master was so annoyed that he sold him to the master of the black girl for whatever the other master was prepared to offer, which seems to have been a good price.

This absurd story was quite probably based on some real exchange that was overheard by the author and then grossly distorted and exaggerated for the sake of theatrical comedy. But such racist portrayals of blacks as ludicrously ignorant, though with some native cunning, are typical of Spanish theater in this slightly later period. It seems likely that such racism had been commonplace in daily life for many centuries.

The most serious problem for Africans in this society was the horrifyingly casual and extreme violence that punctuated everyday life in sixteenth-century cities. For example, in Granada in 1580, some unpleasant banter between two black slaves and two jailors guarding a lockup led to a fight and the death of one of the slaves. Violence was not only a problem for Africans, of course; and in this story the jailors then turned on a white passerby who had stopped to help the slaves, chasing him into a church. There, on holy ground, the two jailors murdered the good Samaritan.

All across Europe, crime rates were rising, and in Spain, Seville had a reputation as the most criminal, vice-ridden, and violent city. Analysis of the limited data available paints a vivid picture of a people ready to kill one another over the slightest dispute. While it may be deplorable, it is at least easy to understand how disagreements over women, money, and business arrangements could get out of hand, leading to hotheaded homicide. But the records also show that men might murder each other over such trivial matters as a seat in a theater, accidentally dirtying someone else's clothing, or a disagreement over the quality of a batch of olives. Nor was such fighting confined to dingy back alleys and poverty-stricken inner-city barrios. Trouble could break out at any moment in any part of the city, by day or night. And these fights quickly escalated into mass battles, brawls which uninvolved bystanders joined, apparently for the fun of it.

As a *SEVILLANA* friend once told me, the devil prefers to strike where faith is strongest, in places where there are many believers. Perhaps she was right, for while sixteenth-century Seville was a place of sin and all forms of human corruption, it was also a center of strong Catholicism and particular devotion to a cult of the Madonna of the Immaculate Conception. This tenet of faith did not achieve acceptance as orthodoxy or dogma until the nineteenth century, but it was supported, sometimes with unholy fervor, in Seville: widespread hooliganism and rioting resulted when a Dominican friar offered a theological refutation of the doctrine during a sermon he preached in the early seventeenth century.

Seville remains a city where, almost uniquely in the modern western world, the rhythm of life is dictated by religion, the church calendar, and the passage of the seasons. Of course, the relationship between church festivals and the progress of the agricultural year is as natural as the relationship between God and life itself. It stretches back to the pagan religions that existed before Judaism, Christianity, and Íslam. But there are few places today where these roots are remembered or appreciated, and nowhere else does a whole city interrupt the ordinary mundane business of day-to-day existence as Seville does at Easter.

During Holy Week, *Semana Santa,* the streets of Seville become a stage on which multiple, breathtaking reenactments of Christ's Passion are celebrated against a backdrop of beautiful buildings. This stage is walked by the Sevillians themselves who take part in their religious rituals as both spectators and actors, but the principal protagonists of this fanfare of collective grief are baroque sculptures, animated only by the spirituality of the spectacle and the artifice of the bearers known as *costaleros* who carry them. It is a most extraordinary experience for anyone lucky enough to be present.

From Palm Sunday to Easter, the eighty official *hermandades* or *cofradías*, religious brotherhoods, set out in procession from the churches and chapels where they are based. Their purpose is to carry the heavy *pasos*, enormous processional floats made of plated metal, intricately worked in baroque designs. The first *paso* always bears an image of Christ crucified, or carrying his cross on the way to Golgotha. Next, there may be *pasos* showing displays of scenes from Christ's life. Finally, the Madonna is borne in outstanding splendor, dressed in sumptuous robes, her *paso* crowned with a glittering canopy on which the devout sprinkle the petals of thousands of flowers, all lit by a thousand candles.

The *costaleros* who toil beneath the *pasos*, hidden from view by a luxuriant velvet skirt, are well drilled and skilled in negotiating the narrow streets. Here, they slow and kneel to slip the Madonna beneath a power cable. There, they stop and turn their burden to face a lone singer who balefully cries a flamenco lament for the suffering of the Lord and the bereft Madonna. In the dark, with the play of the candlelight, the vivid sculptures come to life as these *pasos* move through the crowds of smartly dressed onlookers who squeeze together on the sidewalks.

Walking before and after the *pasos* are the members of the brotherhoods, known as *nazarenos*, dressed in full-length tunics tied with colored cords. They wear long conical masks and carry tall candles, giving them an almost sinister appearance. They proceed through the streets in strictly hierarchical order, and many go barefoot out of penance or devotion. There are brass bands to play a somber march and penitents who carry crosses in imitation of the Savior. All proceed in order, slowly, very slowly, occasionally moving forward, most often waiting patiently, standing silently, contemplating God's sacrifice or thinking about more profane matters.

These brotherhoods are politically powerful and confer enormous kudos and influence on those who govern them; most important, they raise alms for charitable endeavors carried out on a vast scale. Such activities make these brotherhoods an integral part of the social structure of the city as well as custodians of some of the most beautiful baroque sculpture in the world. It is almost impossible for an outsider to understand the extent to which these independent institutions, individually and collectively, influence almost all aspects of life in this city, the provincial capital of Andalusia, a region with eight million inhabitants.

The origins of this enthusiasm among Sevillians for processions at Easter is usually associated with a young sixteenth-century aristocrat, the first Marquis of Tarifa. A year or two before Esteban first set foot in Seville, the marquis returned from a pilgrimage to the Holy Land and Jerusalem, where he had walked in Christ's footsteps, following the stations of the cross. According to tradition, he was awestruck by the spirituality of the experience. He did, however, have the presence of mind to minutely measure the route. On his return to Seville, the pious young man laid out twelve stations of the cross along a route of exactly 1,321 paces (about 1,100 yards). This led from his palace, which has been known ever since as "Pilate's house" (*Casa de Pilatos*) after the Roman procurator who "washed his hands of Jesus," to a cross known as the *Cruz del Campo*, the "Cross in the Country," today rather prosaically the site of Seville's principal brewery.

I was astonished to discover that the *Cruz del Campo* was probably constructed in 1460 by a brotherhood still popularly known as *Los Negritos*, whose members were, until the nineteenth century, almost exclusively black. Moreover, as a consequence, these black African brothers were responsible for the care of one of the twelve stations of the cross established by the Marquis of Tarifa.

It was during this period, when Esteban first arrived in Seville, that *Los Negritos* rose to prominence in the religious life of the city. As a result, it is one of the oldest, if not the oldest, of the brotherhoods that process through the city during Holy Week. However, its origins date back even further still, to a period of terrible famine and civil unrest in the 1390s, when the archbishop of Seville established a brotherhood that would provide moral and spiritual leadership for the especially vulnerable black population.

Perhaps not surprisingly, that antiquity has often been ignored or contested by other *cofradías* disinclined to cede status to such humble bretheren. But *Los Negritos* have always vigorously defended the processional privileges they derive from the great age of their organization. Indeed, they carry their Christ and Madonna to the cathedral on Maundy Thursday, the most important day of Holy Week. That privilege has been maintained over the centuries, thanks to financial sacrifices made by the brothers to pay the cost of litigation against more powerful and wealthier brotherhoods. They have even taken their cases to the highest of all ecclesiastical authorities, the Pope in Rome.

Such determination is easy to explain, according to the brotherhood's most important historian of recent years, Isidoro Moreno. For one of the few ways this disdained social group could assert its ethnic dignity and self-esteem was by affirming its antiquity and its consequent symbolic and ceremonial preeminence within the processions.

In Esteban's day, every Friday during Lent, a procession of *nazarenos* set out slowly for the *Cruz del Campo*, wearing white or black tunics, many of them carrying crosses on their shoulders as a sign of penitence. Some wore hair shirts; others were manacled and carried chains. Some flagellated their bare backs as they processed and others found different instruments of personal torture. Beside them men and women walked solemnly, praying aloud, a lugubrious and sobering sight, characteristic of the time.

In contrast, the return journey was a considerably less pious affair, despite the best efforts of the Franciscan friars to police the occasion. As night fell, given the mixture of the sexes who followed the procession, in the concealing half-light of the burning torches carried by the *nazarenos*, with the penitents all masked and unidentifiable, there were many who again fell into temptation.

This newly fashionable, pious pastime of Esteban's Seville is the origin of the modern Easter processions of *Semana Santa*. It built on a tradition of processions that at the time was focused on the festival of Corpus Christi, when consecrated bread was carried, on display, through the streets, exciting reverence among the faithful. That procession included many flamboyant and eccentric dancers, and there is documentary evidence that during the sixteenth and early seventeenth centuries there were at least twenty-one African dance which participated. One was a group of eight men with tambourines and bells and four women with bells and flags, accompanied by a drummer and guitarist, who represented the battle of Guinea. Spaniards had always embraced the music and dance of Africa. As early as 1451, Africans had performed traditional dances at the wedding of the King's sister. African dance and song had become familiar on the streets of Seville.

But although Africans played a key part in these rituals, that prominence was tempered by the nature of the role that might be reserved for them. Mostly, during the Corpus Christi celebrations they dressed as demons, bearing out the common prejudice that black skin in some way betrayed a blackness of the soul. In fact, it was common practice to sneeze in the presence of black people in order to ward off the devil.

Across Spain and Portugal at that time, black Africans were establishing similar institutions or becoming influential within already existing brotherhoods. In Lisbon, for example, a number of Africans were elected to positions of power in the confraternity of Our Lady of the Rosary. In Jaén, in eastern Andalusia, the processions of one *cofradía* were led by African dancers wearing jingle bells and brightly colored clothes, who moved to the rhythm of a drummer-woman. In Valencia, the main Spanish slave-trading port of the fifteenth century, a "house of blacks" had been set up to as a refuge and source of help and advice to Africans; this institution was eventually given royal status as a brotherhood in 1472.

But these formal organizations were a means of social control the Spaniards imposed on the informal arrangements that originated with Africans themselves. Faced with the growth in the black population and the increasing number of free blacks, the Spanish authorities had established institutions that forced Africans to be more European.

There is a narrow lane in Seville, called the calle del Conde Negro, "the street of the Black Count." It is named after an official title, established in 1475, for a black sheriff appointed to oversee the affairs of the black community. For the first Black Count of Seville, the Catholic Monarchs chose their former servant Juan Valladolid. He was invested in office by a royal decree, which read:

> In return for the many good, loyal and noteworthy services that you have done us and do us each day and because we are aware of your character, ability, and bearing we make you Sheriff and Judge of all the Blacks and Mulattoes, both free and enslaved, in the very noble and very loyal city of Seville and in all its archdiocese. And the said Blacks and Mulattoes shall not celebrate any fiestas except when you the said Juan de Valladolid be present . . . and we order you to investigate disputes and quarrels, marriages and other business that they may have . . . and you know the laws and bylaws must be upheld and we are informed that you are from noble lineage amongst the Blacks.

The appointment of Valladolid and the establishment of a specifically black *cofradía* indicate a society that wanted to give a sense of order to the world. It was the same society that established an inn exclusively for the use of Muslim merchants. The same society also attempted to stop the poor from eating the kind of foodstuffs that were thought appropriate only for the rich, such as game fowl and fruits, a gastro-heirarchy left over from medieval sumptary laws that must have been only imperfectly enforced. And because Seville was the gateway to America, the social and religious attitudes prevalent in the city and moral standards of the *sevillanos* were central to the formation of the society that Spain exported to the New World.

Part Three

WRITING HISTORY

Coat of Arms of Alvar Núñez Cabeza de Vaca (1542). Above the image is an example of a scribal hand characteristic of notarial documents of the mid-sixteenth century. The drawing is part of the case made against Cabeza de Vaca following his failure to govern the River Plate; he was accused of flying his own coat of arms from his ships instead of the royal banner as required. *(Courtesy of Archivo General de Indias; MP Buenos Aires 220)*

9

OVIEDO AND CABEZA DE VACA

THE HISTORIAN'S CRAFT is both the writing of history and the gathering and interpretation of evidence that makes such writing possible. It has been traditional to separate, as much as possible, these two aspects of the process, largely because it is generally desirable to do so for the sake of the reader, and because the detective maintains a sense of mystique by revealing no more than a glimpse of his methods. But from time to time it is necessary for the historian to tell the story of how a particular passage in history has come to be told in a certain way. This usually happens because the passage now needs to be rewritten, as is the case with my retelling of Esteban's tale.

Esteban's American adventure began when Pánfilo Narváez was granted a license by the Spanish king, Charles V, to conquer and settle the vast unknown lands that the Spaniards called Florida. Fortunately, there are two long contemporary accounts of what happened to the Narváez expedition and how Esteban and his three Spanish companions came to be the only survivors out of the 300 men who landed on North American soil in 1528. Those two accounts essentially tell the same story, but they differ in many ways, and many of those differences are significant. In fact, it is possible to write Esteban's personal history only by analyzing those differences, while also looking at other documents that provide facts and details about the expedition. Therefore, before continuing with Esteban's story and the history of what happened to Narváez's expedition to Florida (which will begin again in Part Four), it is necessary to first explain something about the people behind the documentary evidence and the documents themselves.

By far the best-known version of the story, the account which most fascinates scholars and specialists, is a book produced by a commercial publisher in the Spanish city of Zamora, in 1542: *Alvar Núñez Cabeza de Vaca's account of what happened to him in the Indies during the expedition on which Pánfilo de Narváez went as governor.* It tells a sensational story, filled with awesome dangers and miracles that made it instantly popular. Readers soon began to call it *Shipwrecks— Naufragios* in Spanish—and this is the title by which it is usually known today. At the heart of this story is Cabeza de Vaca, who comes across as the hero of every adventure and the savior of his companions. Modern scholars sometimes seem just as bewitched by *Shipwrecks'* improbable tale as Cabeza de Vaca's sixteenth-century audiences were.

Before I read *Shipwrecks* I had the good fortune to come across the adventures of the four survivors in a work that is not nearly so well-known: *The General and Natural History of the Indies,* written by Gonzalo Fernández de Oviedo y Valdés in the 1530s and 1540s. Oviedo, as he is usually known, was the official court chronicler of the Indies, appointed by Charles V to document the Spanish Empire in America.

So whereas Cabeza de Vaca gave an account of his adventures for his own benefit and to make a profit for his publishers, Oviedo had a corporate responsibility as historian royal. The two men wrote for very different reasons and from very different perspectives, and they were also very different in character and temperament.

Oviedo was born into the minor Spanish nobility in 1478, in Madrid, long before it became the capital of Spain. When he was still a child, his parents sent him to live as a servant in an aristocratic household. Strangely, for a man who wrote much about himself and his own worth, Oviedo had almost nothing to say about the mother and father who brought him into the world and so quickly gave him up. That silence has led some scholars to speculate that he may have been illegitimate or perhaps had Jewish ancestry, at a time when Spain was notoriously cruel to Jews and their descendants, even those who had converted to Christianity. But it is more likely that the snobbish royal chronicler preferred to forget the parents who had abandoned him to an uncertain and unloving childhood in the courtly world.

Almost from birth, Oviedo was destined for his long career as an acolyte to the ruling class. He spent his early childhood in the aristocratic household of the Duke of Villahermosa, who recommended the growing boy to the Catholic Monarchs as a suitable companion to Prince Juan, the sickly heir to the Spanish throne. The young Oviedo joined the royal court, a world of aristocracy and privilege that he knew would never be his. But for a time he lived like a prince. He was educated by the most erudite scholars and learned firsthand the ways and intrigues of the court. He also found his place in the world. The institutions of crown, court, and courtier were as parents to the future chronicler of the Indies, providing him material security and the opportunity to serve God and the King. In a sense, his only family was the world of government.

It tells us something of Oviedo's essential warmth of heart that as he matured, this young man, who had suffered such a lack of intimacy and love in youth and childhood, should have fallen in love with his first wife deeply and devoutly. He met Margarita in Madrid, his birthplace, and they were married in about 1507, or perhaps before. He wrote that she was "one of the most beautiful women in the kingdom," and, ever the historian, noted that there were many who could confirm that statement. His description of her beauty first focuses on her hair, which must have been striking. "In truth," he wrote, "I have never in all my life set eyes on a woman with" such locks. Her hair, he said, was "so full and long that she always kept it plaited and doubled over to prevent it from dragging on the ground, for it was more than a hand in length longer than she was tall. Not that she was short," he added, "but of medium build and with the perfect stature for a woman so well proportioned and as beautiful as she." But this "exterior beauty" was the least important of her glories, for she was also virtuous, more so than any of her friends.

"As soon as we were married my dearest Margarita fell pregnant and after nine months she gave birth to a child. But it was a labor that lasted three days and nights, for the baby had died and they had to yank it out. And because only the top of the little creature's head was showing, they had to smash open its skull and remove the brains in order to get a grip. Thus it was born, broken and putrid. And its mother was almost dead." During that awful birth "her mass of golden locks went as white as the margins of the paper" in which Oviedo later wrote, he said. For six months, his beloved hovered on the brink of death, bedridden, paralyzed, and in pain. But she survived to bear him another child, a son, Francisco. Yet in that second birth death took Margarita for her own and left Oviedo to grieve even as he took solace in his newborn heir. Oviedo's life was marked by such brutal highs and lows, swinging from tragedy and tribulation to hope and joy in a continual battering of his emotions. "God lent Margarita to me little more than three years," Oviedo lamented. "But," he concluded, "I cannot do justice in praising her and nor is it pertinent to this history to do so."

Even in an age of brutal childbirth and maternal mortality, it is clear that the young Oviedo's heart was broken by this tragedy. A chance to escape his grief was offered by war with France, and he readily accompanied his sovereign on that campaign. But he went as a bureaucrat, not as a soldier. He then served as secretary to the "Great Captain," his namesake Gonzalo Fernández de Córdoba, the general of a Spanish army sent to fight in Italy. But Oviedo was unhappy as a secretary, and without his beloved Margarita. He determined, at the age of thirty-six, to begin a new life in the New World. He was neither the first nor the last to seek escape from personal tragedy by fleeing to America.

We can only imagine his excitement as he stood on deck as his ship sailed from Spain on April 11, 1514, heading for a continent that had first been discovered when he was only fourteen years old. As a teenage courtier, he himself had seen Columbus beg for royal support and had then seen that support given, in 1492. He then saw Columbus arrive triumphantly at Barcelona to claim his prize as Admiral of the Ocean Sea, with the Atlantic conquered but with almost all America still a blank space on European maps. Now he would see that strange New World for himself.

He now had the title "inspector of mines and slaves." No longer a mere secretary or notary, he was a high-ranking crown official, and he would remain an official of the Crown throughout his years of service in the Indies. He sailed in hope, but would soon be disillusioned by the dishonesty of politics in the colonies, where life was cheap and great wealth was at hand for unscrupulous and self-serving men.

Oviedo opposed those self-serving Spanish captains who were loyal only to themselves. But in those lands, far from court and king, where strength was the law, the priggish, supercilious Oviedo with his aristocratic pretensions must have aroused fury in the breast of every Spanish brigand who hoped to enrich himself. There was an apparent attempt on his life, ordered by the corrupt officials he had angered with his impractical loyalty to the Crown. Oviedo survived, but he had been warned.

Around this time Oviedo escaped these traumas by writing *Claribalte*, a clichéd chivalric romance in the style of *Amadis of Gaul*, a work that is perhaps best left unread, but which has been described as the first American novel—although it has nothing American about it and was simply the first novel to be written in America. In any case Oviedo had become part of the New World, and he would soon embrace that new sense of an American identity in an extraordinary act of official homage to his new home: his *General and Natural History of the Indies*.

Oviedo learned many useful lessons from his fiery American baptism. Most important, he learned that reality in the New World was quite different from the outlandish stories usually told by adventurers when they returned to Spain. In time, he became determined to write a true account of what he had seen. He went to Spain and in 1526 published his *Summary History of the Indies*, a short work, which demonstrated his ability to convincingly document the Spanish experience of America with a strong sense of reality. Charles V was impressed and commissioned Oviedo to write the comprehensive *General and Natural History of the Indies*.

Oviedo combined his role as historian royal with the post of sheriff of the castle at Santo Domingo on Hispaniola, then the administrative center of the Spanish Empire in America. He settled into the comfortable life of a middle-aged bureaucrat and chronicler, documenting everything he could about the New World. But he wrote with the spirit and vision of a pioneer who had experienced great swings of fortune in America and, most importantly, he wrote most of the *History* while he was actually in the New World.

The *History* is a monumental compilation of information and stories. Writing it took Oviedo half his life, and it developed and evolved over time as Oviedo himself changed with age and experience. It was the first attempt ever by a European to give a comprehensive account of the Spanish conquest and exploration and to describe in depth what the Spaniards knew about both North and South America. As a result, Oviedo is rightly remembered by the few who know his work as the "first historian of natural history," the "father of ethnography," and "a High Priest of Truth." But although he wrote the first great study of the New World and may correctly be described as the first European historian of America, Oviedo has been shamefully ignored by all but the most specialized scholars and experts.

Only the first of the three parts of the *History* were printed during his lifetime. For centuries, most of this encyclopedic source of information about sixteenth-century America remained unpublished. The manuscript was filed away, a quietly censored victim of Spanish imperial politics. It was not until the eighteenth century that Spanish historians began the arduous task of preparing his scattered notes and manuscripts for publication. But those historians fought petty campaigns amongst themselves over the bragging rights to write about Spain's history. Oviedo's manuscript was perhaps the most valuable prize in that bureaucratic battle.

Finally, in the 1850s, the Spanish Royal Academy published the first edition. But then, unbelievably, all those papers of incomparable value to the early history of America, which had been so carefully gathered together, were once more inexplicably and disgracefully scattered to the winds of fate. Some parts ended up in Seville, others in Madrid, and still others at the Huntington Library in California; the editor's manuscript, from which that first edition had been typeset, ended up at the Hispanic Society of America in New York.

FOR THAT REASON, one bitter morning in December, I was looking for a bus on Madison Avenue. All about, signs may have wished me "Happy Holidays," but the cold was bruising and no one was in the mood to stop and be helpful. After a while I could no longer feel my numb thighs. I finally found the bus stop and the bus, shaking and shuddering as though it too were shivering with the cold. I sat down by a window and watched us go through Manhattan, all gray and dreary with the winter in spite of the seasonal lights. The bus was busy and crowded at first as we headed north, but by the time we reached 102 Street I was the only passenger. I thought I must have made a mistake. The streets outside also seemed deserted.

Then we swung around a corner and I saw a little group waiting at another bus stop. Within a couple of stops the bus had filled up again, but the passengers were all African-Americans and the whole atmosphere changed. The driver struck up a conversation with a woman he knew and the man she was with, who was in a wheelchair. The gentle banter and chatter of the passengers reminded me of Seville.

But the yuletide spirit and bonhomie were short-lived. As we approached the looming mass of the Cathedral Church of Saint John the Divine, a block or two north of Central Park, the bus once more began to empty. As it climbed toward Morningside Heights, I was again the last one left.

Then the tide turned again. Yuppie students and preppie academics and postgraduates crowded aboard, with their hushed conversations and gravitas. This was Columbia University, and for a few blocks I was reminded of Oxford or Cambridge and the homogenizing purpose of university life. But not for long—as we continued down the long hill to where the A train comes out of the tunnel and goes over a viaduct, I was once more the only person on the bus. And then Puerto Ricans and Domincans began getting on and I began counting the cross streets, keeping an eye out for the Hispanic Society of America.

It is a coincidence that Spanish Harlem expanded up the hill and around the run-down neoclassical grandeur of the Hispanic Society. The building dominates Audubon Terrace, a discreetly monumental plaza flanked by the kind of porticoed stone buildings that appealed to learned societies and museums in the nineteenth century. It was all meant to be the centerpiece of a rich men's suburb, conveniently on the way to Washington Heights. But today the Hispanic Society is instead surrounded by Hispanic immigrants, many of them Hispanic African-Americans.

Audubon Terrace is a strange and unexpected place, slightly sinister. The paltry shadows cast on the piles of graying snow by the half-light of a prolonged December dawn gave it an almost unbearable melancholy. Beyond the double swinging doors of thick glass and solid wood, the hallway was dark, as though still lit by gaslight. Ahead, the courtyard—with its gallery of ancient Spaniards gazing down from their lacquered, or gilded frames—was more peaceful than the medieval monasteries of Spain from which the architectural fabric of the collection had been acquired and brought to New York. It feels like a place of prayer, a labyrinth of the past in which to meditate, but I was briefly puzzled about how far into the past it might really take me. The Hispanic Society had evidently changed little since the nineteenth century. When it opened, it was a brave new world of American scholarship that Prescott, Wright, or Fiske would have entered with wonder, probably awed by its brash, very modern atmosphere of ambition. But today, like Miguel de Cervantes's great literary character Don Quixote, who tried to go back to a life of medieval chivalry, the Hispanic Society of America is stuck in its Victorian past. Also like Don Quixote, it inhabits a distant and almost forgotten corner of the world. As well as being a museum of Spanish artifacts it is also a living reminder of what a museum was like in the nineteenth century; it teaches us how our ancestors tried to understand their past and their history.

To the left of the hallway there is a narrow vestibule with a relatively modern coatrack, perhaps dating from the 1970s. I hung up my coat as instructed by the half-lit concierge, an old man wearing a uniform cut to the dictates of yesteryear. There was an electric bellpush by a wooden door set in the paneling across the end of the room. I pushed the button and heard the bell ring inside.

I was soon admitted to the library. I had traveled here, to an alien corner of a foreign land, so that I could get to know, as best I might, a man who died almost 500 years ago. But that journey was more a pilgrimage than a mission, for I had in fact brought with me a photo-copy of a more reliable manuscript of Oviedo's account of the Narváez expedition the original of which is in the Columbus Library, *La Colombina*, in Seville.

Oviedo had died brokenhearted because he never saw his completed *History* in print, and at the Hispanic Society I was filled with a tragic sense of the past, overwhelmed by my feelings of proximity and distance from the actors who walked the stage of history and the people who wrote about them. With the documents in front of me, I could almost reach out and touch some of that history, but there were too many sights that had to be left unseen, and far too many voices that had to be left unheard. There is a deep melancholy in that incompleteness, a melancholy Oviedo could understand. I was thankful when the day came to an end and I left the Hispanic Society and went back out into the real world again.

In the winter twilight, I began to walk back downtown along Broadway. I had agreed to meet some friends at the steps of the New York Public Library, partly to make a familiar cliché of modern cinema part of my own reality, but also because the library owns one of the very few original, 1542 editions of Cabeza de Vaca's *Shipwrecks*. I had already worked with another of these rare copies at the British Library in London, so my visit to the New York Public Library was capriciously symbolic rather than a scholarly necessity, rather as my time at the Hispanic Society had been. But, as the gloom of evening gathered and I watched downtown Manhattan light up in the night sky beyond Central Park, I realized that strange forces of coincidence and history were at play. It was as though the Fates had symbolically determined that that first edition of the sensational and much published *Shipwrecks* should find a home within that quintessentially popular icon of western learning, the New York Public Library, while, by contrast, Oviedo's scholarly *History*, with its more understated account of the first crossing of North

America, lies almost abandoned in a nearly forgotten library on the fringes of the scholarly world, at 155 Street and Broadway. Nothing could reflect better the way historians have misused these two sources, always preferring the storytelling of *Shipwrecks* to Oviedo's more sober *History*.

THERE IS A moment in Oviedo's *History*—a moment that has gone unnoticed by many scholars—when the Historian Royal passes judgment on Cabeza de Vaca and *Shipwrecks*. Oviedo explained that to write his version of how Esteban and his companions survived the Narváez expedition, he worked closely from a letter or "report sent to the royal officials residing in this city of Santo Domingo by the three noblemen, called Alvar Núñez Cabeza de Vaca, Andrés Dorantes, and Alonso del Castillo. They went with Pánfilo Narváez himself and gave account in written form of what had happened on their journey and where they went." (Esteban is not mentioned—an indication that, as far as the Spanish administration was concerned, he was not considered an important witness to the events.) This report, Oviedo noted, had been mailed from Havana, Cuba, in 1539, where the three Spaniards had stopped off on their way to Spain.

Oviedo then wrote his account of the expedition from the letter of 1539 while he was still in Santo Domingo, probably in 1540 and 1541. He then took the growing manuscript of his *History* to Spain, and in 1547 he met with Cabeza de Vaca in Madrid. After that meeting, and after Oviedo had read *Shipwrecks*, he added an extra, final chapter to his account of the first crossing of America. In effect, that extra chapter is a commentary on *Shipwrecks*.

"I believe," Oviedo concluded in this review, "that the better account is the one written by all three survivors, for it has greater clarity than this other one, *Shipwrecks*, which was written by just one man."

As a conscientious historian, Oviedo would have preferred an account based on the corroborative testimony of three witnesses, as he states. He was no doubt also influenced by the fact that the 1539 letter was the first version of the story that he had read. But he also made it very clear that he was skeptical about *Shipwrecks* because of two definite failings: implausible information supplied with insufficient attention to detail; and, perhaps even more damningly, inaccurate sensationalism.

Oviedo pointed out that in this "later account Cabeza de Vaca claims that where they found the mountains [the Sierra Madre] they also saw samples of gold, kohl, iron, copper, and other metals." But Oviedo goes on to comment sarcastically that, "I would like this explained with much greater clarity and detail." By the time he was writing, Oviedo could be as good as certain that they had found nothing of the sort. He could be quite sure that Cabeza de Vaca was misrepresenting reality.

But Oviedo was even more thoroughly irritated by Cabeza de Vaca's tendency to dramatize his story in a way that turned it into fiction. He was particularly annoyed by an account of an island on the Texas coast (probably Galveston) where some members of the expedition were shipwrecked in 1529.

"I cannot accept that Cabeza de Vaca gives this island the name Misfortune [*Malhado*] in his printed work," Oviedo complained, "because it was not called anything in the first account."

For Oviedo, the invention of the label *Malhado* must have been a particularly irritating example of Cabeza de Vaca's love of embellishment and make-believe. It epitomized for Oviedo the way in which *Shipwrecks* shamelessly perverted history in the interests of a good story. *Malhado* was in fact the name of a fictional island in a best-selling chivalric romance, *Palmerin of Olivia*, a piece of pulp fiction second in popularity only to *Amadis of Gaul*. In *Palmerin of Olivia*, the wicked witch Malfada lived on the island of Malhado, where she turned visiting mariners into wild animals and then kept them prisoner, refusing to allow them to return to their ships. These victims were eventually saved by the hero of the novel, who used his chivalric magic and his supernatural, aristocratic powers to kill Malfada and rescue the unfortunate sailors from their spell.

As far as Oviedo was concerned, there was no place for such fantasies in serious reports about America.

In the case of *Malhado*, Oviedo was doubly troubled. The appearance of a fictional island in a "true" story upset his sense of historical decorum, and it also resulted in a dishonest and therefore innaccurate impression of the Texas Indians on the real island—who, far from casting evil spells over the survivors and turning them into animals, had given them food and shelter.

Cabeza de Vaca "has no right to give" the place "that name," Oviedo stated bluntly. "On the contrary, he himself admits in both accounts that the Christians were well treated by the natives of that island."

Oviedo's antagonism toward Cabeza de Vaca is understandable at a personal level as well as professionally for they were quite different men. As the sun was setting on the Middle Ages, Cabeza de Vaca and Oviedo represented the old and the new, and when they met in Madrid in 1547, in a sense the past and the future of Spain met too.

Oviedo is an archetype of the "new men" needed by the Spanish crown: bureaucrats and administrators, men of government, not men of war. Although Oviedo was a child of the court, his lineage was not exalted, and his early experience made a strong foundation for his views on the social hierarchy. To him, a man's birth was a matter of luck. He considered it madness for a bad man to claim nobility on the basis of lineage, but instead insisted that the root of nobility was individual virtue. Actions, for Oviedo, spoke louder than ancestry. Above all, Oviedo was a servant of Spain, a man who owed everything to his sovereign and the Spanish crown.

Cabeza de Vaca, by contrast, was born into a proud, ancient aristocratic family; his claim to nobility was unquestioned, and in that tradition he was a man of action and a man of the sword. Spanish aristocrats were imbued with a noble tradition of fighting on the frontier with Islam; their medieval forebears had won great estates on the battlefield and had shed their blood to make Spain Christian again. That, at least, was the legend by which the current aristocrats lived, a semifictional history on which they based a code of honor which dominated their public image and which had to be upheld.

Alvar Núñez Cabeza de Vaca was born in the 1490s amongst the sherry cellars and sun-blest wineries of Jerez de la Frontera, the paternal grandson of Pedro de Vera, conquistador of the Canary Islands. But in a period when people often changed and amended their names to suit whatever purpose might please them, he took his even more illustrious maternal surname, Cabeza de Vaca, aligning himself with a widespread and influential noble family of late medieval Spain. Without doubt, he had a pedigree and a noble birthright that would forever be denied Oviedo and his heirs.

Cabeza de Vaca's immediate family was not rich, but as a child he served as a page boy, and later as a soldier in the Italian wars to which Oviedo had gone as a secretary. He then joined the household of Alonso de Guzmán, Duke of Medina Sidonia, head of one of the most powerful aristocratic families in Spain. He served the duke as a steward, a position that brought him responsibility for managing the household finances, and no doubt led him to be appointed treasurer of Narváez's expedition to Florida.

For all his titular power and great wealth, the duke was effete and ineffectual, with a childlike mind trapped in the body of an adult man. In due course, his wife, Anne of Aragon, sued for divorce, saying that he was impotent. At a time when divorce was all but impossible, evidence was required, and so witnesses were called. One of these men testified that he and Cabeza de Vaca had agreed to send "a woman" to the duke's bedchamber. She was a classy kind of call girl, and she "arrived clean, perfumed, and wearing a well-washed night-shirt that gave easy access. . . . They did this," the witness explained, "in order to find out whether the duke was impotent, and the woman promised that she would work hard at the task." Cabeza de Vaca and the witness secretly let the woman into the duke's bedroom and shut the door, but almost immediately they heard their master cry out, "Leave me alone!" Listening intently at the door, the two servants then heard the woman talking lovingly to the duke, praising him, but even so he ordered her to leave. The woman then rushed out complaining that it was all a complete waste of time and she had gotten nowhere. "He is worthless!" she cried. Cabeza de Vaca was involved in two or three further attempts to get the duke interested in women, in the hope that he might learn how to satisfy his wife. But even though these experienced professionals kissed him and caressed him, "holding his member in their hands and trying to stimulate him," it was all to no avail.

Such was the level of intimacy in courtly households of the period.

It is perhaps no surprise that the bookish parvenue Oviedo appears to have personally disliked the aristocratic Cabeza de Vaca, whose adventures in America seemed to outmatch anything Oviedo himself had experienced, and who had usurped the historian's tools, his pen and paper, to write his own account of those adventures, an account which was promptly published, unlike Oviedo's *History*. Oviedo no doubt wrote with considerable personal rancor, but he was also a loyal servant of the empire and he had political motives for finding fault with Cabeza de Vaca's storytelling, for in 1547, Cabeza de Vaca had fallen out of favor with the Spanish crown and was under house arrest in Madrid.

Ten years before, in 1537, Cabeza de Vaca had skillfully wooed his royal audience in Spain with his thrilling tales of North America. He wove such a convincing story that Charles V offered him the governorship of a benighted colony on the River Plate, in modern Argentina and Paraguay. This gift was a poisoned chalice from which Cabeza de Vaca was reluctant to drink, but he knew that success in rescuing that outpost of empire could bring him unimagined glories. He fell into temptation and accepted the challenge. His rule on the River Plate, however, was an unmitigated disaster. The colonists and most of his own men turned on him, imprisoned him, and sent him back to Spain in chains. He had failed as a governor in South America just as he had failed as a conquistador and treasurer in North America. He was no asset to the empire; he was a commander whose men had mutinied. Cabeza de Vaca was now destitute, and Oviedo saw him as simply one more failed conquistador who had not understood the complexity of life in the Americas.

Despite Oviedo's skepticism and despite the obvious embellishments and fantasies that make *Shipwrecks* an exciting story, Cabeza de Vaca's storytelling has captured the hearts and minds of modern scholars and seduced readers ever since it was published. It is a gripping narrative worthy of Hollywood, a quintessentially American story of revenge and redemption familiar from westerns. The usual roles of Spanish conquistador and conquered Indian are reversed with brutal honesty. Now, the European is enslaved and harshly treated. But not for long—for this is also a story of Christian glory, and the four survivors eventually become revered holy men by spreading the gospel of Christ.

The secularizing western world has never been able shake off its cultural and religious heritage, and so Cabeza de Vaca's confessional account of trial by tribulation and hardship that leads to final redemption and a restored sense of order in the peaceful subjection of Indians by Europeans appeals to something deep within the western psyche. The central figure who suffers and is redeemed is Cabeza de Vaca; his sufferings remind us of Christ or the travels of Saint Paul; his experience closely resembles the story of the prodigal son.

As a failed conquistador, Cabeza de Vaca has become an antihero for left-leaning critics, who can use him to assuage their sense of western guilt at the horrors of colonialism and the forging of modern

America. But Cabeza de Vaca was no turncoat who "went native." He did not betray his cultural heritage. He never ceased to be a European, and so his survival is the story of a white man's struggle, the story of a founding father or a pioneer. "Our heroes suffered too," is the cry of liberal pain.

And Cabeza de Vaca claimed that he had used his culture and knowledge to bring the benefits of western morality to a primitive people. He is a missionary, a man of God, of the true God. This image is so Christlike that for almost anyone steeped in modern Judeo-Christian culture he could never seem to be anything but a peace-loving hero. Amongst the chorus of voices convinced of his liberal sanctity there are few dissenters who suggest an alternative image of Cabeza de Vaca.

But Oviedo's skepticism about Cabeza de Vaca's version of events resonated with my own misgivings and doubts that had arisen as I read *Shipwrecks*. The basics of the two accounts were the same, but *Shipwrecks* was filled with surreal and supernatural stories about miraculous shamanism and divine intervention. Dead men stood up and began to walk; bushes burst into flames, providing heat on freezing nights; Cabeza de Vaca performed major surgery with a blunt knife. More than anything else, Cabeza de Vaca seemed to be the hero of every moment, always the center of attention, always the protagonist. Quite simply, I did not believe him.

Then, as I reread Oviedo's *History*, I discovered that although it omitted the more bizarre and implausible miracle stories, it too seemed heavily biased. Andres Dorantes seemed to be Oviedo's hero, always at the heart of the story. I began to wonder why. Why were the two versions different? Why were they different in the ways they were? What might these differences mean?

The origin of the differences between the two versions is relatively easy to explain, for the main storytellers, Cabeza de Vaca and

Dorantes, parted company in 1537 and went on to lead very different lives, so that each had the opportunity to develop his own account to suit his own ends.

Almost as soon as the survivors reached the administrative center at Mexico City, they were required to make the official report to the Audiencia, and Viceroy Mendoza then sent that report to Spain. The following year, Dorantes and Cabeza de Vaca took ship for Spain; but they ended up on different vessels, and while Cabeza de Vaca reached Spain, Dorantes's ship became lost on the coast of the Gulf of Mexico and struggled back to Mexico eight months later.

Cabeza de Vaca took his own copy of Mendoza's report with him and began to tamper with it during the voyage, adding stories of miracles and putting himself at the center of some of the more daring escapades. He no doubt tested these additions on his fellow passengers, to see which of his audacious embellishments would be believed. But he was not fibbing and boasting simply out of vanity; he had a very good reason to do so—he wanted to be appointed governor of Florida himself. These were the first embellishments and changes which Cabeza de Vaca made and which would end up in *Shipwrecks*.

Cabeza de Vaca was to be disappointed, for when he arrived in Spain he discovered that Hernando de Soto had already been appointed governor general of Florida and the preparations for a new expedition to the region were well advanced. His destiny was further disaster on the River Plate. But to ensure the success of Soto's expedition, Charles V and his officials were eager to question Cabeza de Vaca; they wanted as much information about Florida as possible.

In the summer of 1537, unaware that Cabeza de Vaca had already landed in Lisbon and was on his way to court, Charles V ordered his officials in Seville to ensure that "if Cabeza de Vaca had not yet left for [the Royal Court] by such time as you receive this [order] send for him and have him make a sworn, written statement . . . describing everything he saw and learned and heard about the land, settlements, and peoples and then send it to us by the first post."

Clearly the original report that the four survivors made in Mexico City lacked detail. Moreover, we know that once Cabeza de Vaca reached the court, he did indeed supply hundreds of such details, because Oviedo compiled a list of those additions. These were another layer of additions to the story, made by Cabeza de Vaca under interrogation by royal officials, which also found their way into *Shipwrecks*.

When Cabeza de Vaca reached the court, Soto invited him to take part in the expedition, but a Portuguese nobleman commented that he pompously refused "to go under another man's flag." Despite this, Cabeza de Vaca gave a glowing account of Florida, telling all those who asked that although he was sworn to secrecy "he could advise anyone to sell up everything they owned and go there and that by doing so they could not go wrong." Given the experience of the four survivors, one has to wonder whether these reports were not deliberately and maliciously misleading. No doubt the publishers encouraged Cabeza de Vaca to include some of these exaggerations and fictions in *Shipwrecks*, to increase sales.

All this is enough to explain why Cabeza de Vaca was the hero of his own *Shipwrecks* and why it grew to be so different from the original report. He was painting a picture of himself as a tenacious survivor, a commander of Spaniards, and trusted by Indians. But while this explains the differences between *Shipwrecks* and Oviedo's account, it does not account for why Dorantes was so obviously the hero of the latter.

I was to come across the answer to that question thanks to an extraordinary stroke of luck. I discovered a document that convinced me, beyond all reasonable doubt, that Dorantes alone had posted the 1539 letter which Oviedo used as his source for the story. That document also proved that, like Cabeza de Vaca before him, Dorantes was on his way to Spain to ask Charles V for some kind of favor—a job, some land, perhaps a pension. But whereas Cabeza de Vaca had very obviously turned *Shipwrecks* into a story that was almost as much fantasy as fiction, Dorantes's changes and editing had been much more subtle—subtle enough to be overlooked by the assiduous Oviedo.

I avidly read the two versions of the story yet again. I began to read between the lines, not simply comparing the two, but trying to analyze the differences. As I did so, to my astonishment, I realized that in both accounts, whenever anything important happened, Esteban was there. His presence often went unremarked in the accounts, but it could be deduced from other references, which made it impossible that he had been anywhere else. Esteban quickly became an ever-present protagonist, concealed in the silence beneath the noisy words of the Spaniards.

Only then did the implausible stories about medicine men and witch doctors start to make sense. The question I asked myself was this: if it was true that those four survived because they became successful shamans, then which one of the four seems to you to be the most likely candidate for initiating that process?

10

DORANTES AND THE ARCHIVE
OF THE INDIES

THE DOCUMENT WHICH seemed to me to prove that Dorantes must have posted the 1539 letter in Havana, which therefore led me to compare Oviedo's *History* and *Shipwrecks* in new ways and, in turn, revealed Esteban's role beneath the surface of the Spaniards' stories, had been filed away in an archive in Seville, ignored and overlooked by scholars for more than four centuries. The Archive of the Indies, the *Archivo General de Indias*, is the richest source anywhere in the world of original documentary materials for historians of Spanish America. For years, historians had known that some kind of report about Dorantes's military record and his service to the Spanish Empire had been compiled in the sixteenth century, because it was mentioned briefly by his grandson in a book about Mexico published in 1604. But that document had seemed to be lost forever until, one day in 2003, I held it in my hands. For the first time in 430 years someone was about to read the *Dorantes Report* with care and attention.

THE ARCHIVE OF the Indies was established in the late eighteenth century as a repository for all Spanish government papers relating to the empire in the New World and the Far East. No more appropriate city could have been chosen for this institution than Seville, which had been granted a monopoly on all European trade with America by the Catholic Monarchs in 1502. At the same time these monarchs had created an institution, known as the "House of Trade" (*Casa de la Contratación*), to be based in Seville, the purpose of which was to promote and control all of Spain's American commerce. It was this institution more than any other that ordered the production of the manuscript documents that are now stored at the Archive of the Indies in 43,000 separate bundles.

And if the city of Seville was the perfect home for the Archive of the Indies, then no better building could have been chosen to house that institution than the sixteenth-century mercantile exchange, the *Casa Lonja*. This monument to Mammon, imperialism, and Seville's monopoly on American trade was designed by Juan de Herrera, the greatest Spanish architect of his age, who had built Philip II's great monastery-palace of El Escorial in the mountains above Madrid.

A purpose-built exchange was much needed in Seville when work began in the 1580s. Moralists were outraged that the merchants had taken up residence on the *gradas*, the steps that girdled the cathedral. Sermons were preached about Christ and the moneylenders in the temple. But if the merchants paid for the new exchange building in order to save their souls, they also had it built because they were tired of their ad hoc marketplace and wanted somewhere safer and more imposing.

While the Archive of the Indies still stores its documents in the exchange building, ready to yield up their tiny fragments of the historical mosaic, the reading room has recently been more practically rehoused in an eighteenth-century palace across a small side street. The steps at the entrance to this new building give a breathtaking view of the historical context. Ahead is the cathedral, rising in ever decreasing curves of squat but elegant flying buttresses, roofs, lanterns, and Gothic pinnacles. To the left is the high, solid wall of a city within a city, the Alcazar, "erstwhile palace of the Moorish kings, richly and beautifully decorated in the Arab style, with sumptuous marble everywhere. Water flows through channels constructed with artful ingenuity, through its many baths and rooms. It is, indeed, a very pleasant place to pass the summer," the sixteenth-century Italian ambassador remembered. "There is a courtyard full of well cared for orange trees and lemons and beyond there are extensive gardens of great beauty. It is thickly planted with oranges that offer shade from the sun's rays, so that it may indeed be the most pleasant place in all Spain." The Alcazar continues to offer the solace of peace and quiet, and shaded glades where cooling water trickles in the high heat of summer, and it remains a welcome oasis for all sorts of tourists, not just ambassadors.

The long-windowed rooms of the Archive of the Indies, with their high ceilings and blinds, and the narrowness of the Arab-built streets outside, all contribute to a calm and ageless sense of shade. This is a place of contemplation, where scholars from around the globe slowly meditate on the possible meaning of the manuscripts on the desks before them.

The search for historical treasure is quietly paced to the rhythm of reading the handwriting of a different age. In the sixteenth century, professional scribes often scrawled their words and used a personal shorthand of strange abbreviations, so that their script seems like Arabic or Persian to the untrained eye. Names were written down as they sounded, changing their spelling from scribe to scribe. Even common words were spelled differently by different scribes and notaries. Murmuring lips betray the scholar's need to read aloud in order to make sense of a problematic passage in a difficult document. And it was thus that the men and women of the past to whom these scholars devote their researches read these very same documents, always pronouncing the words out loud.

Every document holds some unknown detail about the past, but it is rare to find previously unseen material about important events like

Pánfilo Narváez's expedition to Florida. Yet for centuries, scholars have been on the lookout for such material.

For many decades now, any search for obscure documents was very laborious. Such a search began with the handwritten catalog, and any papers that looked promising had to be ordered from the staff. Then there was a long wait, while they were brought up from the storerooms. But recently there have been changes at the archive, including the scanning of tens of thousands of documents. The scanned images can be read at computer terminals in the reading rooms, printed out, and taken home. More recently still, many have become available on the Internet. Most important of all, the vast catalog, the work of generations of librarians and archivists, filled with detailed and extensive entries for every document in the collection, has been made available electronically.

In 2003, I spent two months in Seville searching this miraculous catalog for the names of all the individuals I could find mentioned in the context of Narváez's expedition. I was partly interested in whether anyone besides Esteban, Castillo, Cabeza de Vaca, and Dorantes had survived the overland march, but I found no trace of others. I was more interested, however, in what had happened to the men and women who stayed with Narváez's ships after he and his army had set out into the Florida swamps. Thanks to the electronic catalog, I was able to verify that many of those men and women had survived, but at great personal cost. Some ended up in Mexico City, desperately poor and willing to turn their hand to anything, and many went on to be involved in the brutal campaign in northwest Mexico, men like Aguayo and the Castañedas mentioned in the second chapter. Esteban and his Spanish companions came across the aftermath of those campaigns when they first made contact with the Spanish slavers, as described at the very beginning of this book.

By searching for the name "Dorantes" and carefully examining every one of the sixty-five catalog entries and documents highlighted by the search engine, I came across the document that contained the crucial clue to the fact that Andrés Dorantes posted the 1539 letter: the *Report and Proof of the Service and Merits of Baltasar Dorantes de Carranza*. It had been produced at the instigation of Andrés Dorantes's son Baltasar, in the 1570s, which is probably why it had been overlooked by historians for so long, for it appeared to be too late to contain anything much about Andrés Dorantes and was anyway apparently about the son, not the father.

Despite the late date, far from being about Baltasar Dorantes, on close inspection, the *Dorantes Report* seemed to be mostly about Andrés, his father, survivor of the Florida expedition and Esteban's legal owner.

The sixteenth-century handwriting was difficult to read in places; and so, because I wanted to use my limited time in Seville to continue my search for other new material, I arranged for the scanned images of the document to be printed off and I filed the copies away to work on when I was back in London. But weeks later, when I turned to the copies again, I found that whoever scanned the documents had missed a crucial page: what was missing was most of the testimony given by a witness who seemed to confirm that Dorantes had in fact been to Spain and further testimony that appeared to explain what had happened to him after he arrived in Mexico City. I was furious. But not for long, because there was only one solution: this was a wonderful reason to go back to Seville the following spring.

IN LATE APRIL of the following year I arrived to find the city filled with the scent of jasmine flowers, honeysuckle, and orange blossom. It was early evening, just as the chill of night was taking the heat off the late spring day. I dropped my bags at home and set out into the warm promise of a Seville night. I soon found my friends in a small square not far from the archive, sitting under a lemon tree, drinking beer, and playing music. With them was a human rights lawyer from

New Mexico called Jim Anaya who had come to spend a sabbatical at the Seville law school. As we shook hands for the first time, I sensed that Jim would become part of my search for Esteban. Years later, he would come with me on the final leg of my quest, when we went together in search of Esteban's death, but that story will have to wait until its proper place at the end of this book.

Jim and the *sevillanos* were warming up for the annual fiesta that takes over the life of the city for a week in April every year. The Dorantes document, I decided, could wait for another week. It had, after all, already been waiting for nearly four and a half centuries.

As a doctoral student I lived not far from the precinct on the outskirts of the modern city that is reserved for Seville's extraordinary April fair, the *Feria de Abril,* a fiesta quite unlike any other. The *Feria* takes place on a valuable tract of real estate that lies abandoned for fifty-one weeks of the year but which blossoms into a vast party, a carnival of music, color, pageant and social pretension for the other seven days.

Living nearby, I was fascinated by the bizarre sight of thousands of lavatory pans beginning to sprout, day by day, on the *Feria* ground, lining up in disorderly ranks, as though the first shoots of spring. These were soon encased in jerry-built cubicles, rickety restrooms for the partygoers crammed into 1500 giant marquees of colored woven canvas. Alternately red-and-white or green-and-white, these marquees sit squat, side by side, raised above cobbled streets and sidewalks of dark ochre sand called *albero,* the same unique sand from near Seville which soaks up the blood of men and animals spilled in the bullring.

Nearly all these marquees are privately owned, and for a week toward the end of April they are filled with beautiful Andalusian girls dressed in bright polka-dotted flamenco frocks, transported there in exquisitely restored nineteenth-century carriages. Smartly dressed riders in wide-brimmed hats, on perfectly groomed mounts, steer through the traffic jam of antiques, pulling up here and there at the curb. The polka-dotted girls bring them sherry and snacks and then gracefully ride sidesaddle behind their beaux.

The air is filled with a chaotic counterpoint of flamenco rhythms, clapped out by a million pairs of hands. Inside the marquees, small crowds gather, some with guitars, some with rattles made of split bamboo. There is sometimes a deep bass drum to mark the rhythm. The formal dancing begins.

But *Feria* is more than a beautiful bacchanal, it is a also a social and cultural exchange. This is as much a place of work as a playground, a vast forum in which to make new friends and acquaintances. It is a place where ideas and information are bartered amid the banter and the music.

At the *Feria*, two years before, an idle conversation about my work on *Shipwrecks* led to the unexpected discovery that two friends of mine, Carmen and Teresa, are both related to the Cabeza de Vaca family. Such coincidences are a useful reminder that history is simply the extension of the present back into the past and that time is an intangible concept measured only by the ephemeral ticking of a clock. Alvar Núñez Cabeza de Vaca's genes, much diluted, stood before me, drinking sherry and eating prawns, just as surely as Esteban had once walked the streets around the Archive of the Indies.

At that moment Carmen's husband, Luis, gave me the information that explains this digression in my narrative away from matters of the past into the apparently unrelated Seville *Feria*. Luis told me about the recent improvements at the Archive of the Indies: "They've now finished the computerization," he explained. "Everyone says its marvelous."

Over the years, our minds play with memories, and the heat and sherry of the relentless *fiesta* blurs one *Feria* with another into a sense of timelessness. This is a part of life that is usually on hold, waiting to be resumed for a few days in every year. It is a parallel universe. Now, as the midnight fireworks sounded the temporary end of *Feria* until the following year, I said good-bye to old friends and new, including Jim Anaya, and the following morning, back on real time, I turned my attention to the *Dorantes Report*.

PHYSICALLY, the *Dorantes Report* consists of twenty-three folios of heavy sixteenth-century paper. The notary wrote on both sides, in handwriting typical of the period. The black ink has turned brown with age and in some places it has burned through the paper. It has a scent unique to such documents, something between the dilapidated smell of a fused electrical circuit and that promising scent of a newly opened cigarette packet. It is tied together with a white ribbon, forming a bundle with various other similar documents, all stored in a somewhat collapsed archive file the size of a small fruit crate, beige in color and marked on the spine in printed characters "Mexico 212."

What was this "report and proof" for? Why did Baltasar go to the trouble and expense of making an official record of his father's deeds twenty years or more after Andrés Dorantes had died?

To answer these questions we need to return to what we know of Spanish history in the Middle Ages, before the discovery of America in 1492. During the eight centuries of the reconquest, when the Christians in the north slowly defeated the Muslim rulers of Spain, town by town, province by province, the Christians divided the conquered lands into many small kingdoms and principalities, each ruled by its own sovereign and each with its own aristocracy.

The kings and princes themselves led their armies of aristocrats into battle, and in the aftermath of victory they personally rewarded their men. That close personal relationship between the sovereign and subject is well illustrated by the much-romanticized myth of how the Cabeza de Vaca family rose to nobility, a story recorded under the entry for *Vaca* in a Spanish dictionary that was published in 1611.

In the summer of 1212, Alfonso VIII, King of Castile, was preparing for a massive and decisive confrontation with the Muslim south then ruled by the Almohad dynasty. The battle known as Las Navas de Tolosa took place on July 16 and has gone down in history as the pivotal moment in the Christian reconquest. But the victory was not easily won. The Almohad forces had marched high into the Sierra Morena in order to take control of the mountain passes, blocking the Christian progress south into the rich valley of the Guadalquivir River.

As King Alfonso consulted with his generals, a young shepherd, called Martín Alhaja, whose surname means "something valuable" in Arabic, was admitted to the royal quarters. He claimed to know of a secret pass through the mountains, at a place he was used to taking his sheep. He is said to have marked the crucial turn on the trail with a cow's skull, enabling Alfonso to outflank the Almohad army and thus win the day for Christendom. When the humble Alhaja was ennobled as a reward for his service, he took the surname *Cabeza de Vaca*, "cow's head," which became the emblematic and striking image, so appropriate to Texas, featured on Alvar Núñez's coat of arms.

Given the central role of the Lord as a "shepherd" in Christian iconography, this image of a shepherd leading Christian soldiers to a great victory is no doubt too good to be true. The story is presumably a legend demonstrating God's intervention on the side of Ferdinand III. But this legend perfectly illustrates the way in which subjects had direct, personal access to their king.

For those who had served their kingdom well, in peace or in war, it was practical to go to court and approach their ruler and beg of him some reward or favor. During the medieval reconquest of Spain from the Moors, those who fought valiantly and successfully were usually rewarded with land, peasants, and a title of nobility. The relationship between the ruler and the individual was personal. But as Spain and her empire grew, that personal relationship became more difficult.

Wars were fought on many fronts, kings and princes often did not take command of their armies, and many subjects lived their whole lives without ever seeing their ruler.

When Charles V came to the throne in 1517, he was also the prince of many kingdoms in Italy, Germany, the Netherlands, and France. He was king of all Spain and ruler of all the Spanish possessions in America. His captains and generals were soon to reach Asia by way of the Pacific Ocean. It had become impossible for those who served him throughout the empire to beg favors of him personally. What is more, his servants in the farthest reaches of the empire were beyond his direct personal control. Central government had become as remote from those it claimed to govern as it has perhaps ever been in the history of humanity.

As with the black conquistador of Mexico, Juan Garrido, the Spanish crown solved this problem through recourse to the Audiencia, requiring the viceroy and magistrates to oversee a "hearing" at which an individual could present his case for some kind of royal reward. That is precisely what Baltasar Dorantes did in the winter of 1573 to 1574.

Baltasar was now twenty-five years old, and his father had been dead for ten or fifteen years. He had married young and against his family's wishes and now found that he could not support his wife and five children. So he turned to the Spanish crown for support, requesting a salaried post within the burgeoning bureaucracy necessary to maintain at least some degree of royal control over the empire. His petition took a standard form, and the *Dorantes Report* in the Archive of the Indies is one of tens of thousands of such documents produced by the network of imperial *audiencias* and other similar institutions. First, Baltasar presented, almost certainly orally, a *memorial* or *información*, in which he explained his personal needs in some detail and gave an extensive account of why he deserved some royal reward. Next, the court prepared a series of questions based on this introductory speech. These questions were then put to a series of witnesses, friends and acquaintances of Baltasar. A transcript of that oral process was made during the hearing, and fair copies produced afterward. The document in Seville is one of those copies.

Baltasar himself seems to have achieved little in his twenty-five years, beyond a hungry family that he was unable to feed. But in a society that was obsessed with ancestry and lineage, it was logical and natural for a son or even a grandson to claim a reward for his father's or grandfather's achievements. So Baltasar set about finding witnesses who might attest to Andrés Dorantes's "services and merits."

The document in the archive is an official notary's copy of the transcript of the "hearing" in the Audiencia of Mexico, which was sent to the crown officials in Spain. Inundated by such dossiers about unknown subjects in remote lands, the crown officials no doubt sent Baltasar's *Report* to the royal archive at Simancas, where it was filed away unread. There it remained until the Archive of the Indies was established in the late eighteenth century and all such documents relating to America were sent to Seville. It next saw daylight, or lamplight, when some diligent nineteenth-century archivist glanced at it, created a catalog entry, and returned it to the bundle of documents alongside which it nestled, undisturbed, until it was recently scanned.

THE *DORANTES REPORT* only becomes meaningful when it is read in context, so that the detailed information it contains can be added to other information we have about Andrés Dorantes's life. For me, it has been very important for two closely related reasons. Later, I will describe how it helps to explain why Esteban was chosen to lead Marcos de Niza's expedition to search for the Seven Cities of Gold. But it also supports the contention that Dorantes went to Spain in 1539 alone, and that he alone posted the account of the Florida expedition which Oviedo relied on to write his *History*.

First, *Shipwrecks* reports that in 1537, a year after the four survivors had arrived in Mexico, Andrés Dorantes and Cabeza de Vaca decided to take ship for Spain. But while they were waiting at the Mexican port of Veracruz to set sail, Cabeza de Vaca realized that there was something wrong with their ship and switched to a different vessel at the last moment. There is then no record in *Shipwrecks* that Dorantes ever made it to Spain; and on the basis of that information, it has always been assumed that Dorantes never went to Spain again, despite the fact that the Cosmographer Royal, Alonso de Santa Cruz, had reported that Cabeza de Vaca and Dorantes were seen at court together. However, the *Dorantes Report* explains what happened in 1537 in much greater detail.

Thirty-one years later, on December 7, 1573, an old and illiterate sailor called Sebastian Granado testified that he had first met Andrés Dorantes about thirty-five years earlier, when Dorantes first arrived in Mexico from Florida. Granado remembered that "sometime later" he and Andrés Dorantes had sailed for Spain together aboard a ship captained by Sancho de Piniga and added that Dorantes was on his way to Spain to beg favor of the king.

For some reason, which Granado does not record, their ship failed to reach the Atlantic, and instead Captain Piniga sailed about the Mexican Gulf and the Caribbean for eight months, apparently lost. During this time, Granado remembered, Dorantes had kept the sailors and passengers royally entertained with accounts of his adventures in Florida. Evidently, as he told those tales, Dorantes began to develop his own version of the story, testing it out on that captive audience.

Granado himself offered no explanation for Piniga's strange or incompetent captaincy, but a letter written to the Spanish crown by Viceroy Mendoza in December 1537 records something of what happened. Piniga's ship eventually reached port at Tabasco, on the Yucután Peninsula in Mexico, where he attempted to illegally unload gold, silver, and other goods belonging to the Crown. Piniga was then arrested and jailed, but it is not clear whether he had planned the theft from the outset, or whether this was the opportunistic act of a desperate man who had been lost at sea for many months.

The innocent passengers were allowed to go free by the authorities at Tabasco, and they made their way back to the main port of Veracruz, presumably hoping to resume their intended voyage on other ships.

Mendoza also stated in his letter that as soon as he learned that Piniga's ship had not reached Spain, he ordered Dorantes to come immediately to Mexico. "When he arrived," the viceroy reported to Charles V, "I spoke to him and informed him that he would do well to return to those lands" through which he had traveled when he survived the Florida expedition. Dorantes was to take with him "some priests and friars and cavalry officers appointed by me, in order to explore and find out precisely what was there." Mendoza emphasized what a great service such a mission would be to both God and King. Dorantes bowed to this pressure, replying that "he would be delighted to do so." Mendoza went on to estimate the cost of the expedition at 3,500 or 4,000 pesos, which he intended to pay out of taxes due to the Crown. And Mendoza later reported to Charles V that after many discussions with Dorantes, he had spent a considerable amount of money on an expedition to include forty or fifty cavalrymen.

We can only speculate as to why Dorantes changed his mind about leading an expedition to the north between the beginning of 1537 when he sailed for Spain and the end of the year. Mendoza may have bribed him. The months he had spent lost at sea may have deterred him from further voyages. But whatever the reason, by the beginning of 1538 Dorantes was clearly preparing to lead Mendoza's proposed expedition in search of the Seven Cities of Gold.

Yet Dorantes never went, and Mendoza's letters to the Crown offer absolutely no explanation for what went wrong. The viceroy simply stated that he had no idea why the whole business had failed. But one

of Baltasar's witnesses, Andrés Dorantes's stepson, Pedro de Benevides, knew only too well. He told the Audiencia, in 1573, that Mendoza
had indeed at first "chosen Andrés Dorantes to go as captain of the
cavalry company he intended to send to the said lands," the Seven Cities of Gold. But, then, "a Franciscan priest," a "Frenchman called
Marcos de Niza," had somehow convinced Mendoza that under no
circumstances should he send a military expedition. Instead, Marcos
de Niza offered to go himself, taking a few Franciscan friars with
him, and he asked that the viceroy arrange for him take Esteban as a
guide.

Looking back across the centuries, we can see the hand of the powerful Archbishop of Mexico, Juan de Zumárraga, at work, for we know
that he had put heavy pressure on Mendoza to leave responsibility for
staffing the expedition to the Church. Charles V agreed with Zumárraga and so, in the fall of 1538, Esteban and Marcos de Niza left for
the Seven Cities without Andrés Dorantes, on an expressly peaceful
and religious mission to evangelize the Indians of the north.

With his prestigious captaincy revoked, Andrés Dorantes no doubt
turned his back on Mexico in disgust and prepared at once to sail again
for Spain in search of some more reliable reward that might be granted
him directly by the emperor. The *Dorantes Report* contains very good
evidence that he made that voyage: three of the witnesses state that
Dorantes made a successful journey to Spain and that he returned to
Mexico with a royal warrant granting him lordship over two estates.
The problem is that those witnesses do not give the dates of that trip.

So, when did Dorantes go to Spain?

We can arrive at a convincing answer to this question through a
simple process of deduction. First, we know that Dorantes was still in
Mexico in August 1538, because his signature appears on Mexican legal documents with that date. Second, we also know that Dorantes
helped Mendoza suppress an Indian rebellion in Mexico in 1542, and
to take part in that campaign he must have been back in Mexico by

the winter of 1541 to 1542. Therefore, Dorantes must have traveled to Spain between August 1538 and the early fall of 1541. In fact, a royal decree dated December 1539 refers to an account given to the King by Dorantes relating to his stepdaughter's inheritance. Although the document does not make clear whether this account was given on paper or in person, the fact that Dorantes had business with the Crown that year is reason to think that he may well have been present.

We also know that Oviedo said the source he used to write his *History* was sent to him, in 1539, from Havana, which was the usual last port of call in America before the Atlantic crossing for ships sailing from Mexico to Spain. Although Oviedo claimed that letter was posted by all three Spanish survivors of the Narváez expedition, we know that this was impossible. First of all, Cabeza de Vaca was in Spain throughout 1539. Meanwhile, there is no evidence that Castillo went to Spain until, possibly, he made a trip sometime between February and November of 1541. So, by a simple process of elimination, if one of the three posted the letter, then it was almost certainly Dorantes.

All the evidence suggests that the 1539 letter, which Oviedo used as his source, was sent by Dorantes alone, who was on his way to Spain to beg favor of his king and sort out his stepdaugher's inheritance.

Dorantes comes across as the protagonist of Oviedo's account, and now we know why.

Dorantes had three years during which he had edited and reworked his own copy of the original report given to the Audiencia in 1536. He had exactly the same motive for doing so as Cabeza de Vaca had for creating the story told in *Shipwrecks*.

The letter posted by Dorantes in Cuba in 1539 was clearly closely based on the original report; otherwise, Oviedo would never have described it as a letter sent by all of the survivors. Therefore, we can conclude the changes made by Dorantes were understated by comparison with *Shipwrecks*, which had been heavily reworked in Spain, where unlikely stories about America seemed more believable to an audience with no experience of the New World. *Shipwrecks* was also the work of the publishers in Zamora, interested in sales as much as plausibility. By contrast, Dorantes's unspectacular editing made his account seem all the more plausible because there was so much less emphasis on the marvelous, the weird, and the wonderful. As sleight of hand, it was considerably more sophisticated, sufficiently so to deceive Oviedo.

In one way and another, *Shipwrecks* was largely rewritten by Cabeza de Vaca in order to present himself in the best possible light to Charles V when he arrived at court in 1537. It was, in effect, a job application—and a successful one, for he was shortly afterward appointed governor of the River Plate. In exactly the same way, the report Dorantes sent from Havana to Santo Domingo in 1539, on which Oviedo based his version of the story, was intended to serve Dorantes's interests. He was not appointed governor of a remote colony, but he did return to Mexico a relatively rich and privileged man. He too benefited from putting himself at the center of the story he told.

There is perhaps one example among many that best illustrates this crucial difference between Oviedo and *Shipwrecks*. That is the way the two books record one particular encounter between the four survivors and a tribe of Indians, which must have taken place somewhere in northeastern Mexico in 1535.

Both accounts report that when a group of Indians who had been accompanying the survivors on their westward journey refused to continue, the survivors became very angry.

Oviedo records that Dorantes then told the Indians that as a result of their intransigence they would all die. Almost immediately, some kind of epidemic struck their terrified Indian guides and many died, while many more fell gravely ill.

Shipwrecks, by contrast, tells of how Cabeza de Vaca left the camp and went to spend the night away from the others, which terrified the

Indians. A frightened delegation of Indians then sought him out and pleaded with him not to be angry, saying they would go anywhere the survivors wanted them to. Then, "the strangest thing happened: the very same day many of them fell sick and the following day eight men died."

The incident was crucial to the four men's survival and their eventual return to Spanish Mexico, because it greatly reinforced the Indians' perception of them as magicians and godlike figures. It may have simply been a coincidence that some European disease, perhaps carried by the survivors, struck the Indians at this moment. Or the epidemic may have been caused by some mass psychosomatic response, as some commentators have suggested. In fact, the survivors themselves seem to have been doubtful about the true connection between the cunning subterfuge of their theatrical anger and the subsequent malaise that struck their Indian companions.

What really happened is of no concern to us at the moment. But it is of critical importance that Dorantes appears to have taken the initiative in cowing the recalcitrant Indians in Oviedo's *History*, whereas Cabeza de Vaca plays that role in *Shipwrecks*. For, if each of these noble Spaniards could claim for himself the credit that was claimed by the other, then there is really no evidence at all that the credit was not in fact due to Esteban or Castillo.

The narrators of both *Shipwrecks* and Oviedo's source were unreliable, especially when it came to the way they reported themselves and their own actions. Their accounts can neither be dismissed as fiction nor trusted as fact. Faced with such a dilemma, the conscientious historian can only read between the lines, digging below the surface for telltale clues within these texts and in other documents that suggest what really happened and who the real protagonists were.

In that respect, the biggest clue of all is surely the assertion in *Shipwrecks* that it was the black African slave, Esteban, who "always spoke to the Indians, gathering all the information they needed"——and that Esteban "found out about the trails that they wanted to follow and what nations, tribes, and settlements might be thereabouts." It seems from those remarks almost as though Esteban may have been the leader of this group of wanderers, or at least that he played the pivotal role in their survival.

And, on reflection, this would hardly be surprising. After all, which of the four survivors had the greatest breadth and depth of experience amongst different societies and religions? Which of these four men had traveled most widely and known the widest variety of men? Which was best prepared mentally to adapt to the ways of an alien culture? Who among them had already survived great cruelties and privations?

Part Four

THE EXPEDITION TO FLORIDA

1527–1536

The Rio Grande above Big Bend, Texas, with author in foreground. Esteban and the three Spaniards came across a scattering of cultivated fields in the valley bottom in this area, a welcome sight, as these were the first signs of agriculture they saw during their crossing of America. (*Courtesy of Elizabeth Jones*)

II

ATLANTIC CROSSING

1527

ON NARVÁEZ'S ORDERS, the town criers announced the coming expedition and drummers beat out the rat-a-tat-tat of their recruiting drums in the towns and counties of Spain.

America seemed a blank page on which a determined Spaniard could write almost any ambition into history. The wonders of Mexico were now widely known. The news of Hernan Cortés's astonishing conquest had been quickly spread by the publication of his letters to the king, letters glittering with tales of gold, silver, and adventure. They were published almost as soon as they arrived in Spain, rushed into print at the great publishing house of Seville run by Jacob Cromberger. In north America, the first European captain to set his eyes on Florida, Juan Ponce de León, had once claimed the Fountain of Eternal Youth might be found somewhere deep within modern Georgia. Florida is so called because it was first sighted by a European in 1513 soon after *pascua florida,* "flowering Sunday," as Easter Sunday was known because of the flowers associated with the Resurrection of Jesus Christ. From the moment it was christened it became a land imbued with strong symbolic significance, a place of resurrection, redemption, and desperate suffering.

For hopeful conquistadors, America always seemed an empty space on their European maps, waiting to be filled with wealth and adventure. For all that a few cautious and experienced men like Oviedo warned that the blank was likely to turn out a land of disaster, the hopeful adventurers took no notice, their imagination filled with personal glory. Among those who first set eyes on Florida from Ponce de León's ship had been Juan Garrido, African conquistador and citizen of Mexico. His presence at that first sighting seems almost a portent for the recurrent prominence of Africans in the very early accounts of the history of Spanish Florida.

During early 1527, at Seville, deep within the cool aisles of the cathedral, captains and merchants struck their deals and prayed for safe returns, both human and material. Meanwhile, outside, America was as much a fiction as a fact when you listened to the stories bellowed forth by Narváez's agents from the terraced *gradas* of Seville, where soldiers of fortune and all kinds of riffraff waited for a chance to sail the Ocean Sea and make their dreams come true.

Oviedo was clear about the true nature of such men, the agents of Spain's enterprise in the Indies. Far away in America he had seen many good men lost and many villains make their fortunes. He warned his friends and countrymen: "Should you decide to come to the Indies, you must at least, while still in Seville, find out everything you can and be certain that whichever captain you may sail with is the kind of man who can fulfill the promises he makes. Take heed of what is offered, for you are lending him your life and entrusting yourself to his will. For many of these captains promise rewards that are not theirs to give, for they are ignorant, they have no knowledge of the New World. They pay for your very being by buying you with words that carry less weight than a feather, for at least when a feather is carried on the wind and you can see that it is blown around without direction, you know it has some physical existence. But the words of a liar have no substance and once spoken they are invisible and disappear like thin air. And even if you have some contract signed and sealed by a bank or marked up in the official records in Barcelona, is that going to make you believe what you are told by some captain who comes to America to discover what he has never seen and knows nothing about?"

But Narváez's gullible and foolhardy followers had a captain who had himself eschewed these worldly words of caution. Oviedo commented that were Narváez to have looked back on how badly his expedition to arrest Cortés had turned out, he would not have risked more trouble by sailing for Florida. But Oviedo knew in his heart that men like Cortés and Narváez were driven by envy and bitter rivalry. They may have sought salvation in the conversion of heathen souls, and they may have become wealthy by exploiting and stealing from the Indians, but they were driven to explore and conquer unknown America simply "because it was there," as George Mallory famously said of climbing Everest, shortly before he lost his life there.

The *gradas* and the nearby drinking houses that lined the narrow, foul-smelling streets leading to the waterside were the newsrooms of the New World. In those crucible taverns of alchemy, wine was the catalyst that quickly turned rumor, fact, and fantasy into a fool's gold of greed, ambition, and a desperate desire for adventure.

As Narváez's expedition slowly came together, the hopeful gentlemen and younger nobles, who were poor if not penniless, but were rich in conceit and arrogance, begged, borrowed, and pawned their possessions to pay their passage and buy horses, weapons, and slaves. Few records of these transactions have survived, but from those that have we know that on March 29, 1527, two months before Narváez set sail on his ill-fated journey to perdition, one of the expeditionaries, called Pedro Lunel, was given permission to take four black slaves to Florida. Then on April 12, García de Paredes and Diego de Solis were each granted leave to carry four black slaves as well. From this limited evidence we can infer that many more took black slaves with them, and that Esteban was but one among many. But almost all of them, masters and slaves, were doomed, destined to die anonymously on those strands afar remote, where no drum would be heard, nor a funeral note, as their cadavers rotted in the tropical sun.

Narváez's ships left Seville in May, making their slow descent of the treacherous Guadalquivir River, the aqueous artery of Andalusia, with its shallows and wrecks ready to bring disaster to any poorly piloted ship. Some of the company were already aboard, guarding the limited cargo that had been loaded at Seville. But most made their way by road. They were headed for the Atlantic port of Sanlúcar de Barrameda, at the mouth of that great river, from which Columbus had set sail in August 1492.

There were rations to be bought to sustain the 600 men and women aboard the five ships during the month or longer that the crossing might take. Each man and woman aboard could look forward to a daily allowance of a couple of pints of water and a couple of pints of wine. The contract may have stated clearly that this must be "good wine" and perhaps specified that it must be "from the Manzanilla region and not from the riverside vineyards. Nor should it be sour, nor woody in flavor, but instead it must be of the *fino* sherry type, with a good nose and palate." The factor or steward was to judge the quality, but such men were easily corrupted and standard fare on board seems to have been universally poor.

Many passengers brought their own food, but on an expedition such as Narváez's most of the men had to put up with what they were served. Typically, in addition to half a pound or less of hardtack biscuit, each man would get no more than a few ounces of beans or chickpeas and a quarter pound of salted fish on Mondays, Wednesdays, Fridays, and Saturdays. On Tuesdays, the men were treated to an ounce or two of rice, cooked with oil and half a pound of salt pork or *tocino*, cured pig fat. And on Sundays and Thursdays there was salt beef or mutton and an ounce or two of cheese. So little water and so much salt food were a thirsty combination.

All this food and drink, the passengers, the crew, their personal possessions, their weapons, the ship's cannon, the ballast, and any other cargo all had to be packed aboard along with the horses and their fodder and an assortment of chickens and other livestock. One English traveler recalled the raucous response of crewmen and passengers when their "captain bought three Spanish hogs: the roughness of the weather made them so sea-sick that no man could forbear laughing to see them go reeling and spewing about the decks." Then, of course, there were the rats, the mice, and the fleas. At times it must have seemed like some hellish interpretation of Noah's Ark.

The captains allowed those rich enough to afford some privacy on the high seas to pay the carpenters to build them coffinlike cabins, stacked together on the forecastle. There was barely space to lie cramped in these rickety slots, but still their occupants filled them with jars of water and other foodstuffs. Belowdecks, life would be even more unpleasant. The inhabitable space of the average ship was the size of a large urban apartment, about 1,800 square feet. It carried 100 to 120 men and women.

THE CROWDED CHAOS of Sanlúcar as the great fleets were readied for the crossing to America can be readily imagined nowadays, when, at Whitsun every year, in late May or early June, the town fills with pilgrim-revelers. Thousands throng the beaches, waiting to be carried into the forests across the mouth of the stream. From the far shore, they will make their way by foot, horse, covered wagon, or tractor to the holy shrine at El Rocío.

To the modern observer, this peculiar Spanish tradition, like Holy Week in Seville, seems like a journey back in time. The crowds in the streets and those waiting on the beaches sing and dance a kind of flamenco as they set out on a great mobile party that, over five days of drinking and eating and music-making, will slowly snake its way among the dunes and squat Mediterranean pines of the last great shoreline wilderness in western Europe.

At the break of the day, with sun glinting on the mist and cloud rising from the water, the mouth of the river fills with boats, while on the left bank, the early-morning pilgrims breakfast on the shrimp for which the waters of that coast are famous, and the dry sherry wine called Manzanilla, that tastes of the sea—cold, and crisp, and salty. A counterpoint to the chattering and excited voices of those about to embark on an adventure rings through the streets, cut by the quick clip-clop of horses and the rumbling of multicolored ox-drawn wagons.

This annual pilgrimage takes place in defiance of modernity, one of the only authentic, almost medieval gatherings to be found anywhere in the developed world.

It is a precious clue to what Narváez's men must have felt as they took ship for the unknown, for as one waits in line on the banks of the Guadalquivir for the chance to cross the water, there is an overwhelming expectation of the unexpected. Despite the coming discomforts and hardships of the pilgrimage, there are hopes, there are fears, and there is a great feeling of freedom, a libertine liberty perhaps, but freedom nonetheless that comes with crossing the threshold of a new and different world.

The pilgrimage of El Rocío is merely a fortnight-long party, albeit of great spiritual intent. It holds none of the terrors and dangers faced by Narváez and his men. It can only hint at their trepidation as they too packed the seaside town of Sánlucar, gorging themselves on prawns and sherry, eating fresh meat, and filling their bellies to bursting for the last time before setting out for America.

ON JUNE 17, 1527, Narváez's five ships left Sanlúcar and, just as Columbus had done, set sail for the Canary Islands, for that was the best way to pick up the westerly trade winds and their surest route to the New World and the conquest of Florida. There were 600 men and a handful of women aboard these boats, among them our principal storytellers: Captain Andrés Dorantes and the treasurer Alvar Núñez Cabeza de Vaca. Also aboard was the young Alonso del Castillo Maldonado, the doctor's son from Salamanca, still a teenager. And also perhaps Esteban, who, as he sailed down the African coast, must have watched for the shore with strong emotions. These were his last glimpses of the continent where he was born. It was a sad farewell, but he must also have felt the thrill of adventure as he headed for the New World and the promised land the Spaniards called Florida.

In the official records the voyage to Santo Domingo passed without incident, but all long voyages were dangerous and unpleasant. In letters home, a sixteenth-century passenger, Eugenio de Salazar, recited the refrain of all transatlantic travelers, "To sail the sea is to count its dangers," and he likened the ocean to the Styx and his ship's captain to Charon, the gatekeeper of Hades, ancient ferryman of the dead. There were storms to be feared, but equally terrifying was the possibility of being becalmed and baking in the noonday sun with no sign of a breeze. The crowding and seasickness coupled with the lack of toilets and washing arrangements made for squalid, unsanitary conditions. Many died of disease.

Atlantic voyages had become such a cause for concern by 1536 that Archbishop Zumárraga complained to the Council of the Indies that his very soul was suffering at news recently arrived in Mexico City from the port of Veracruz. The flotilla from Spain that had brought the royal mail had been devastatingly struck by illness and over a third of the passengers had died. On the road from Veracruz and in Mexico itself more than 200 had perished in the inns or by the wayside. It was a terrible tragedy, and not uncommon. Perhaps it is only surprising that not all ships making the Atlantic crossing were similarly afflicted, given the miserable existence aboard.

As Narváez's ships hit the open seas, the roll and pitch began to claim its victims. A friar who accompanied Bartolomé de las Casas during one difficult Atlantic crossing remembered that the sea had

quickly proven to be no place for a human being. All the friars were
struck down with such seasickness that they were like dead and dying
men. They could not move about the ship, and some lay so incapaci-
tated that they needed the old hand Las Casas to bring them chamber
pots and basins to be sick into.

It was an experience etched into the hearts, minds, and stomachs
of everyone who sailed to the New World and survived. Salazar recalled how he squeezed with his family into the tiny coffinlike cabin,
"three palms high and five palms square." And there they succumbed
to the power of the sea, which unsettled their balance and their bellies
with such belligerence, that all aboard, parents and children, the old
and the young, with the pallor of death on them, began to feel nausea
in their very souls. They cried out "*Baac, baac, baac*" and "*Bor, bor,
bor,*" and then, as one, their stomachs poured forth each and every
morsel they had consumed that day and the day before.

Salazar had suffered the first days of sickness with his eyes shut,
holed up in his stinking cabin. But in due course, he heard a prayer: he
does not make clear whether it was the imagined muttering of his own
befuddled mind or the words of the ship's pastor. He now turned to
his wife and remarked that though he had been convinced that their
ship was home to the devil himself, he had now heard the words of
God and it was time to rise from his sickbed, see the sea, and find out
what was happening up on the deck. Salazar had found his sea legs.

"And so," he remembered, "I readied myself as best I could and escaped that whale's stomach of a cabin in which we were confined." On
deck he found that they had taken ship in the kind of vessel "that is
variously referred to as a hobby horse, a wooden nag, or a flying pig,"
but which Salazar thought to be more like a town or city, "although
not exactly the City of God so piously described by Saint Augustine,"
for there was no church, no law court, no ordained priest, "nor any of
the other usual signs of civilization." Physically, this civic space was
long and thin toward the front and wider at the back; it had its squares

and streets, its accommodation, and its city walls. It had its water source, he remembered, called the pumps, which "produced a liquid abhorred by nose and palate, and which even seems repellant to the eye. For it comes bubbling from the depths of hell and reeks like the devil's breath."

In fact, working the pumps was a terrifying as well as a revolting business. A chalk line was drawn inside the hull, indicating the point above which the foul liquid that collected in the depths of the vessel must not rise. For, were it to do so, the ship would go down and that meant certain death for all. This put those who worked the pumps under a terrible strain psychologically, while they endured perhaps the most unpleasant working conditions imaginable.

Salazar likened the masts to trees that, instead of perfuming the city streets with fragrant blossoms, polluted the atmosphere with the smell of rotting caulking. Warming to his theme, he went on to write of "terrible rivers of unimaginable filth on which float seasick lice that having gorged themselves on the ship's boys, then vomit great chunks of human flesh."

Underfoot, the streets of Salazar's floating town were slippery when it rained, and sticky and gooey when dried by the afternoon sun.

But there was plenty of exhilarating hunting for those so inclined, with an abundance of scuttling cockroaches to test your aim and rats that, like wild boar, turn and face down their pursuers to test one's nerve.

Mealtimes were a disgusting frenzy of repulsive food and unseemly greed. As the sun set, the page boys would come up from below decks, carrying filthy rags that they referred to as tablecloths. They then piled up crumbs of hardtack biscuit, which looked like little mounds of dung, and set out three or four wooden platters replete with the evening meal of cow bones, without their marrow, but with the odd badly cooked sinew still attached.

"The board is set, the board is set!" one of the pages cried out.

"Captain of the Ship, Sir! Master Mate, Ship's Company, the board is set! The food is ready. Long live the King, long live the King of Castile, King of land and sea! May his enemies be beheaded and no wine for any dissenters. All haste to the table!"

Salazar now turned his sarcasm on the diners themselves. "In the time it takes to say Amen the sailors sat on the floor around their table, some with their legs stretched out, some bent under them, some squatting. Then, without waiting for grace to be said, these Knights of the Round Table brought out all sorts of knives. Knives for slaughtering swine, others for cutting the throats of lambs, and some for cutting purses. With these they would set about dissecting the sinew-laden bones as though they had studied surgery at the great Spanish universities."

And while you did your best to eat this fine fare, Salazar complained, you were constantly jostled by a crowd of men and women. The young and old, the filthy and the clean, were all crammed up next to each other, some burping, some puking, some farting, and others shitting themselves, all as you were trying to have your breakfast.

"And," he went on, "if you need to do your business yourself, my friend, you will have to hang your body out over the sea and make your offerings to the sun, the moon, and the planets. And mind you hang on to the mane of that wooden nag, for if you fail to do so, it will throw you and you will never ride again. And it is such a terrifying business that many a turd has shown its face, only to retreat back up its arsehole like a turtle into its shell, all for the fear of falling into the sea."

But if a transatlantic crossing was mostly endless days of dull routine and unpalatable food eaten in uncongenial surroundings, there were moments of respite and relaxation. Card games were a constant concern to the crown authorities because passengers and crew often ran up frightening debts that led to fighting and even murder. But the captains connived at such gambling, for they or their stewards charged rent for the packs of cards.

There were occasional cockfights and evenings of guitar music and song when the weather was good. There was gossip and conversation, and some spent their time writing. But one of the most popular pastimes was to gather around while some literate passenger read aloud from one of the popular books of the day. There can be little doubt that as Narváez's expedition sailed the Ocean Sea, men such as Cabeza de Vaca and Andrés Dorantes read from works such as the best-selling novel *Amadis of Gaul*.

In 1571, a passenger called Cristóbal de Maldonado was confined belowdecks by the captain of the *Santiago* for complaining vociferously that the women aboard were fornicating with just about everyone. These harlots, he was horrified to report to the amused captain, paraded about the deck, brushing up against anyone, and even getting involved with the apprentice seamen. Maldonado, either too mean, too prudish, or too wary of disease, was of course clearly suffering as a result of his own unsatisfied urges. As with many a prude before and since, his moral censure seems to have been the result of envy.

Other examples of libidinous excess conform to the stereotype of the sailor who seeks relief with a cabin boy. In 1591, the Indies fleet was anchored near port when a bedraggled thirteen-year-old page boy was spotted struggling in the water and hauled aboard the flagship. He had, he claimed, fled from a steward called Caravallo, who had kissed him and tried to mount him.

The steward was arrested, and a number of page boys gave evidence that he had been pestering them for some time. His usual method was to lure them into the food store with a promise of a "mouthful" and then lock them in. A boy called Pedro Merino said that on one occasion, when this happened, Caravallo began to kiss him all over his body. He then began to touch his buttocks and grope him elsewhere. "Suspecting he was a sodomite," the boy reported, he had decided "to sleep with his trousers well tied." The boy remembered that on another occasion, he woke in the night to find Caravallo kissing him, saying that he wanted to have sex with him. Astonishingly, Merino simply hoisted up his trousers and went back to sleep. Soon afterward, the boy again awoke, this time to find Caravallo rubbing his "nature" up against his backside.

The investigators also interviewed a seventeen-year-old Basque apprentice, who had gone into an enclosed space with Caravallo in order to compare the size of their members. But the Basque said that when he was propositioned by Caravallo, he refused, and added that "if Caravallo wanted to release his passions . . . then there was a ram aboard ship with which to do it!"

Caravallo was tied with his wrists behind him and was then hauled up slowly on the yardarm, dislocating the joints in his arms and shoulders. Weights were tied to his feet. Seven times they raised him, each time with more weight, until he foamed at the mouth and puked. Although he would not confess, he was condemned to receive 300 lashes. But he was to be spared by the Fates, for he appealed to the House of Trade and was sent to Seville. On arrival, he disappeared and never faced the deadly justice that early modern Spain imposed on men who committed "the sin against nature."

The words of contemporary travelers give the modern reader the impression that this floating inferno was more like the world of low-life picaresque fiction than the real world experienced by ordinary men and women. But America was a land of the unimaginable, a place of dreams and nightmares, and the passage there across the Ocean Sea was itself a gateway into the wholly strange otherness of that brave new world. To sail the Atlantic was to surrender oneself to an alien, magical reality, often at once horrific and beautiful.

That great crossing began in earnest once the ships left the last islands of the Old World, the Canaries, ten days or so out from Sanlúcar. After a few days of respite while provisions and water casks were replenished, the passengers entered the world of fiction. Whether thirst and hunger, the hot sun, the driving rain, and the fear of storms turned their minds or whether the New World was a truly miraculous place is hard to fathom. But that most sober of commentators, Oviedo, devoted a whole chapter of his *History* to an account of all the strange and almost supernatural places he heard about in America and the many magically real events that Spaniards had experienced there.

For most voyagers, that strange and alien world began at sea. On one seventeenth-century crossing, as the Marquis of Villena's ships set out across the Atlantic, leaving the Canaries to stern, a young woman called Esperanza gave birth to a baby girl. Esperanza was married to one of the marquis's closest aides, and mother and child were as well cared for as the shipboard conditions permitted. But the mother fell ill, and her breasts would not produce milk to feed the suckling infant. For eight days the child was kept alive with lemon curd mixed with almond milk, but then she too fell ill. The bishop of Neuva Vizcaya was aboard and baptized the ailing infant, for the doctors had concluded that she would not live to see sunset.

It took a divine miracle to save the girl. As the desperate father pleaded with the doctors to find some remedy, a hunting dog, a bitch, entered the cramped cabin. This dog was replete with mother's milk, for she had borne a litter only a week before. The desperate father gathered up the dog and held her canine teats to the lips of his dying child. For three weeks, the dog clambered daily into the child's crib, until the ship reached Puerto Rico. But as the chronicler of this unlikely tale remarked, it is easier to believe in Valerius Maxiumus's story of Romulus and Remus, the founders of Rome who were suckled by a she-wolf, than to give credit to such a contemporary miracle!

NARVÁEZ'S CROSSING IN 1527 was disturbed by the terrible fears of one of the women passengers. There is a strange chapter in *Shipwrecks*, possibly written by Castillo, which tells an extraordinary story. Aboard the governor general's ship, one woman foretold the many disasters awaiting the expedition. She predicted that few if any of those who marched inland from their landing place on the Florida coast would live to see their loved ones again. Any who might survive, she said, would be blessed by God, who would work great miracles through them.

Narváez replied with a speech to rouse his men. "We will fight and we will conquer," he predicted. "We will see many strange peoples and pass through strange lands. Many will die in those conquests, but for the victors there will be riches and glory."

"But how," the governor general wanted to know, "had this hysterical woman come by such pessimistic predictions for the outcome of his expedition?"

The woman explained that, while still in Spain, an old Moorish hag from the town of Hornachos had foretold the whole business. This witchlike character had cast a doom-laden spell over the woman adventuress who sailed with Narváez—who, in turn, seems to have extended that spell to the crowded company aboard ship.

With hindsight, *Shipwrecks* presents this soothsaying heathen woman in a way that suggests supernatural clarity and accuracy of prescience, while all but calling the shipboard wench a whore-mistress. As was mentioned in Chapter 2, as soon as the expedition had landed on the shores of Florida and some of the men marched inland in search of glory, this protesting woman persuaded the other women to "forsake their husbands" and take up with the men who stayed on the ships.

The story of the Moorish woman whose dread philosophy was preached upon the high seas by an apparently hysterical Siren seems strange and surreal. But there are many strange passages in *Shipwrecks* which describe events that seem marvelous and supernatural and which assert an image of America stranger than a fairy tale. For the sober historian in a rational age, when magic is properly considered hocus pocus, merely trickery or sleight of hand, the work of a conjuror, such accounts of the strange and otherworldly inevitably sound a warning bell. But in the sixteenth century, when every sober, rational monarch believed unquestioningly in God and Purgatory, when kings and princes surrounded themselves with augurs and astrologers, such pessimistic prescience was neither fact nor fiction, but lore, to be respected if not believed outright.

With cabin fever eating away at the sanity of even the most sanguine aboard ship, the sight of the Antilles was a welcome relief. Ennui now gave way to impatience as the whole company longed for their arrival at Santo Domingo, the Spanish capital of the Caribbean. But the Moorish woman's predictions would soon haunt the expedition.

12

CUBA

1527–1528

THE EXPEDITION SPENT forty-five days at Santo Domingo arranging provisions and, in particular, buying the horses they would eventually be forced to eat on the shores of Alabama. During that time, 140 men deserted. Captains and commanders were accustomed to unscrupulous emigrants who would hitch a ride with a military expedition and then desert when they landed in their promised American New World, but by all accounts more men than might have been expected chose to jump ship at this juncture.

These deserters were no doubt in part persuaded by the gloomy predictions of the Moorish soothsayer and the repeated incantation of those predictions aboard ship. Also, Narváez's men can hardly have been encouraged by local reports of Florida and the north coast of the Mexican Gulf.

A century and a half later, the geographical territory covered by the southern states still remained almost entirely Indian country. When a French expedition led by the indefatigable René-Robert Cavelier de La Salle touched at Santo Domingo on its way to establish a colony at the mouth of the Mississippi, the picture painted by buccaneers, pirates, and other old hands of the New World was so sordid and evil that many of La Salle's Frenchmen deserted. They preferred piracy to the horrible privations of a world of alligators, ferocious insects, and fearsome Indians. There is little reason to think that the motives of the deserters from the Narváez expedition were any different.

The intrepid English traveler Robert Tomson suggests that by the 1550s, Santo Domingo remained a moderately prosperous port of about 500 Spanish households. Although he complained about the cassava bread, which he did not like, he reported finding plenty of beef, mutton, and cows and noted that there were wealthy farmers with more than 1,000 head of cattle. He mentioned the "good store of hogs flesh," which he described as so "sweet and savory and so wholesome that they give it to sick folks to eat instead of hens and capons." Even today, Spanish cured hams made from the small Iberian black pigs fed on acorns remain one of the great gastronomic pleasures of the world.

At Santo Domingo, the local mosquitoes had their first taste of Narváez's men, and the men had their first experience of tropical mosquitoes. Tomson offers an exemplary novice's account of the experience:

The country is for the most part of the year very hot and very full of a kind of fly or gnat with a long bill that bites and troubles people very much at night when they are asleep. It stings their faces and hands and other parts of their bodies that are uncovered, making them swell up amazingly.

For Tomson, the horrors of the mosquitoes were perhaps equaled by another pest:

There is also another kind of small worm, which creeps into the soles of men's feet and especially the blacks and children, who are usually barefoot; and it makes their feet grow to the size of a man's head and ache enough to drive one mad. They have no remedy for this other than to cut open the flesh, sometimes as much as 3 or 4 inches, in order to dig them out.

Santo Domingo held only the toughest of attractions for the deserters, because the colony had become depopulated and had fallen on hard times. Interest in the Spanish Caribbean islands had waned in favor of exploration of the American mainland. One deserter, Francisco Díaz, tried to settle, but there was nothing to hold the attention of a man in search of glory and he soon sought his fortune in Colombia.

Santo Domingo was also a terrifyingly expensive place, where merchants easily marked up their prices by 200 percent and their customers readily paid. But the crown forbade the local government to impose price controls, for only such vast rewards would persuade merchants to run the risk of ruin involved in trading with so isolated a colony. Pirates lurked offshore, patrolling the seas in search of ill-defended merchant ships, and there were storms that would sink a ship without a trace.

Even the roads that ran out of town to the hinterland farms were dangerous: travelers were menaced by robber bands of Indians and escaped slaves. The priests were few and reluctant to travel, so many a lonely rancher died unconfessed and unblessed. Those priests and friars who dared to serve their flock were often corrupt and unscrupulous, cajoling the sick and dying into making wills in favor of the parishes and monasteries.

WITH MORE OF his men abandoning the expedition for the false promise that Santo Domingo held for the gullible, Narváez was anxious to move the remaining loyal men and his fleet to his personal stronghold, Cuba. No man would dare desert Governor General Pánfilo de Narváez on Cuba, the island where his wife, María de

Valenzuela, ran the vast estates he had won with the cut and thrust of his own steel. On Cuba he was a powerful oligarch and the inhabitants, Spaniard and Indian alike, well remembered Narváez's brutal campaigns against the Indians, so vividly described by Las Casas.

Narváez's barbarity had brought fleeting wealth to the first conquistadors in Cuba, but by 1527 the island was in sharp decline. Too many Indians had been murdered or had died from disease; those who remained to work the Spaniards' farms were the living dead who had lost their natural will to live, to love and breed, to sing and dance, to resist their captors. They had no strength to work the mines and fields that made their captors rich.

But Narváez was still rich and powerful in that wasted land of broken people. The governor general, with the treasurer Cabeza de Vaca at his side, now concluded the purchase of another ship at Santo Domingo, and by the end of August they had moved the fleet to the Cuban port of Santiago. Narváez was temporarily home.

They now urgently needed to replace the deserters. Narváez knew that he would find some men with experience of Florida who might be prepared to join his great venture. For many years, the colonists on the Caribbean islands had quietly and secretively carried on an illegal trade in Indians captured and enslaved on the northern coasts of the Mexican Gulf. There can be little doubt that some such skullduggery was in Narváez's mind as he prepared his expedition, for he had long played his part in that illicit business.

The late summer of 1527 was an especially propitious time for Narváez to be on the lookout for old Florida hands. There were many men milling around the ports of Santo Domingo, Cuba, and nearby Puerto Rico who, the year before, had escaped with nothing but their lives from a disastrous attempt to settle the Atlantic coast of modern Georgia. They had sailed with the most expensive colonizing fleet in Spanish imperial history, led by Lucas Vázquez de Ayllón, a long-standing enemy of Narváez, who lost his own life at the settlement he established briefly at Sapelo Sound. The few men who survived returned to Cuba and Puerto Rico completely ruined. Many of these destitute refugees must have joined Narváez's expedition. We can only identify one of them, Antonio de Aguayo, who was among the small group of men who were to survive the Florida disaster because they stayed with the ships and who ended up in Guadalajara with Nuño de Guzmán. It also seems likely that a man called Pedro de Valdivieso, a cousin of Andrés Dorantes, also survived Ayllón's colony. He was a veteran of the conquest of Mexico and Cortés's subsequent mission to Honduras, which had also ended in disaster. He too joined Narváez's expedition to Florida and eventually lost his life on the shores of Texas.

I HAVE A tentative theory that Esteban had also escaped Ayllón's colony and that Dorantes first met Esteban in Cuba through Valdivieso. We know that there were many slaves among the men and women who abandoned Ayllón's hopeless settlement. It is even perhaps plausible that Valdivieso sold Esteban to his cousin in order to pay for his passage with Narváez. Although it is usually accepted that Esteban sailed from Spain with Narváez's fleet, there is no actual record that he did.

Lucas Vázquez de Ayllón was a corrupt, ambitious lawyer and administrator who had become rich in the New World by financing slaving raids in the Bahamas. But as the islands were stripped of their people, those expeditions became fruitless. His captains soon needed to look beyond the outlying Caribbean islands, and in due course they started to explore the Atlantic coast of Florida.

Early expeditions took some slaves, but also reported the promise of rich agricultural lands, which led Spaniards to describe the place as a "New Andalusia," reminiscent of the fertile fields of southern Spain. Ayllón seized his opportunity, for the Spanish Crown was then promoting peaceful settlement and trade with the Indians.

Florida would provide the ideal opportunity to establish such a colony. And if a little quiet slaving went on unheeded by the Crown, well, what of it?

Ayllón went to Spain to plead his case and was richly rewarded with an exclusive contract to explore further and to settle that newly discovered land.

Oviedo, with his acerbic hindsight, wrote that while he well believed that Ayllón was a good administrator and an able judge, "which was why he studied law," he was not well suited for conquests. "He who would lead soldiers, should himself a soldier be," the royal historian remarked. It is Oviedo who tells us the sorry tale of Ayllón's colony.

The problems began as soon as the expedition reached the site of the proposed settlement at the mouth of the south Santee River. The flagship sank, taking most of the supplies down with it. Worse was to follow, for it quickly became clear that this was no place to found their colony, and further exploration led Ayllón to move the party south, to Sapelo Sound.

It was a cold autumn and the expedition was now short of supplies. To compound their troubles, the men seem to have soon polluted the freshwater supply where they had pitched their camp, and disease struck mercilessly. The commander fell sick and, on his own saint's day, October 18, the day of Saint Luke, Lucas Vázquez de Ayllón died.

The expedition now split into two factions, and as they fought among themselves, the Indians attacked, killing many of the Spaniards. That fraught situation came to a head when two characters whom Oviedo portrays as upstart rebel leaders imprisoned some noblemen who remonstrated with them.

The rebels now plotted to murder those who had dared oppose their proletarian rule. But on the very night that they determined "to put those evil plans into effect, a group of black Africans set fire to the rebel leader's house," where the noblemen were being held captive. These slaves were motivated by their own revulsion at the rebels, or so

Oviedo implies, indicating that they rose up on their own initiative. As the leading modern scholar of this incident, Paul Hoffman, had pointed out, the slaves were probably mostly household servants rather than agricultural laborers. There must have been "extreme personal provocation" for such individuals to have taken sides in a dispute between their masters.

The timely intervention of these slaves, as Oviedo interpreted the story, led to some semblance of order in the colony. The incarcerated noblemen escaped and overpowered the rebels, who were then, in their turn, imprisoned. The decision was now taken to evacuate the remains of the colony, and the refugees made their way back to the Caribbean.

So many of the colonists had perished on that hostile shore that many of the slaves who had saved the colony from the rebels now found themselves in the unusual position of having no master. Some presumably considered remaining where they were, but the increasing appearance of aggressive Indians and the godforsaken conditions they had so far experienced seem to have decided them to return to Spanish jurisdiction. On their return, they once again submitted to being chattels, now owned by the beneficiaries of their masters' wills or by their masters' creditors. In many cases, the surviving slaves may have been almost the only legacy left by the deceased Spaniards.

Many of those slaves would therefore have found themselves put up for sale by executors, widows, or children as they attempted to pay off the heavy debts incurred by the colonists before they set out. With their experience of Florida and their apparent record of loyalty, these slaves must have seemed an especially attractive purchase to Narváez's men. One day someone may be able to prove that Esteban was among them.

As NARVÁEZ SET about rebuilding his expedition on Cuba, he ordered Cabeza de Vaca to take two ships to the town of Trinidad where one of his cronies, a local official, had offered to supply the expedition with the contents of a warehouse or store he owned. As soon as they anchored in the poorly sheltered bay that served as a harbor for Trinidad, the pilots became uneasy. Foul weather was brewing and they were anxious to find a more secure haven for the ships.

The following morning, the skies had darkened and it began to rain so hard that the men given leave to go ashore went back to their ships to be out of the cold and the wet. Had they taken refuge in the town, they might have lived long enough to perish alongside their companions on the shores of Alabama and Texas.

A canoe now appeared, struggling through the waves, bringing word to Cabeza de Vaca that he should go to the town and take official receipt of the promised supplies of food and other necessary items. But Cabeza de Vaca preferred to stay aboard the ships, or so he claimed in a letter to the Crown. By midday, the messengers had returned with another missive, begging the treasurer to go to the town. They had also brought him a horse to ride. It was providential and proof of Cabeza de Vaca's luck, for although he again said that he preferred to remain with the ships, the pilots and sailors urged him to complete the business as soon as possible so that they might seek a better port in which to see out the coming storm. Those old sea dogs knew what was coming and they were terrified that the ships would be lost if they remained where they were.

Cabeza de Vaca set out on the short journey to La Trinidad, but the vicious wind now whipped up the sea. The terrified mariners were trapped, unable to launch the longboats and unable to beach the ships. "By Sunday night," Cabeza de Vaca recalled, "the rains and the tempest raged such that it was as stormy in the town as it was out to sea. All the houses and the church collapsed and we were only able to stay upright and walk about when seven or eight of us grabbed a-hold of one another."

All night they huddled together, unable to find respite from the storm. "As we walked in the woods, we were as fearful of the falling trees as we were of the collapsing houses in the town." And "all night, amongst the great thunderclaps, we could hear the murmur of voices and the sound of bells and flutes and drums and other instruments."

Cabeza de Vaca and his companions shivered and shook until daybreak, when the storm relented and the tropical sun once again shone in a blue sky. There was no sign of the ships at the port, although the search party saw their buoys, a sure sign that both vessels were lost. Half a mile inland they found one of the ships' boats, swaying in the brisk breeze, caught in the canopy of the trees. Thirty miles down the coast, two of the mariners were found smashed against the rocks, "so disfigured by their injuries that they were unrecognizable." This is the earliest known description of a hurricane.

The survivors remained at Trinidad for five days, the supplies spoiled and town destroyed. They were rescued by Narváez, whose ships had survived the storm. With the sense of foreboding caused by the loss of two ships, sixty comrades, twenty horses, and so many valuable provisions, the expedition set sail to spend the winter at the good port of Jagua.

Those few weeks of winter were the last repose the expedition was to enjoy. The Cuban hurricane had been the first of an increasingly relentless cataract of disasters that scythed through Narváez's men just as his own band of brigands had cut down so many defenseless Cubans. God was against them, and even as Narváez's ships crept out of the safety of Jagua and again took to the high seas, they were headed for immediate misfortune because of the bad luck of their pilot, Miruelo.

Miruelo seems to have had some experience sailing the northern coast of the Mexican gulf, but misfortune and mishap seem to have been his constant companions, as though he had been cursed by the evil eye. "A pilot is to a ship what the soul is the human form," according to Alonso de Chaves's sixteenth-century Spanish seamen's manual, the *Mariner's Mirror*. A pilot needed to know the currents, the great geographer Velasco wrote in 1571, "which serve to retard or accelerate the voyage. For, when voyaging where the currents flow, vessels can drift many leagues and travel much more than seems to be the case. Sometimes, indeed, the currents have such force that they will carry a ship against a contrary wind."

But Miruelo was evidently unlucky and has been accused of incompetence, and as they sailed for Havana, he grounded the fleet on the shoals of Canarreo off a deserted and remote stretch of Cuban coast. For a full fortnight the ships lay trapped by the exposed reefs, with their keels grazing the sand, until another storm blew in and flooded the shallows long enough for them to reach the high seas. But the Fates were again unkind, and as Miruelo tried to enter Havana harbor, a powerful south wind blew them far out into the Mexican Gulf.

13

FLORIDA

1528

WITH EASTER FAST approaching, the clouds receded, the storm relented, and the expedition landed they knew not where. There is a consensus among historians that their landfall was a few miles north of the entrance to Tampa Bay, but Miruelo and the other pilots evidently had no idea of their location. Some have suggested that the sailors thought they had landed somewhere on the west side of the Gulf of Mexico, near Pánuco in northeast Mexico. But that is difficult to believe, for however incompetent these pilots may have been, they must have known that the sun moves from east to west; therefore, they must have realized they were on the west coast of the Florida peninsula, because that is the only coastline in the Gulf of Mexico where the sea is to the west and land is to the east.

Paul Hoffman offers the most convincing explanation of what happened to them, which to a great extent exonerates Miruelo from his reputation for ineptitude. Hoffman argues that as soon as land was sighted, Miruelo and the other pilots agreed that the expedition should seek a good harbor known as Bahía Honda, the "Deep Bay" (Tampa Bay), which they knew was somewhere nearby. But there was a crucial difference of opinion. Most of the pilots had no personal knowledge of that coast and had to rely on a written description by

Alonso de Chaves. Miruelo, on the other hand, knew the coast well and realized that Chaves was wrong and had placed the harbor too far to the north.

At first, the other pilots took their readings and all agreed that the harbor was to be found to the south. But as they coasted in that direction, they passed the latitude at which Chaves had marked Tampa Bay. Miruelo insisted they continue, but the other pilots protested. There was a standoff. Hoffman argues that when Narváez found himself faced with this impasse, he ordered Miruelo to continue the search to the south alone, perhaps partly to get him out of the way.

The remaining pilots now became utterly confused, ignorant of the coast on which they were stranded and unable to determine where they were. Their problems were compounded by the fact that during the early Spanish exploration of the Gulf of Mexico, Tampa Bay in Florida and Matagorda Bay in Texas were both sometimes marked on maps as the "Bay" or "River of the Holy Spirit." The pilots now managed to conflate the two and measured the length of the gulf coast without taking into account all of Alabama, Mississippi, and Louisiana, and much of Florida and Texas. That mistake was deadly.

Even Narváez eventually lost patience with his helpless pilots. On Easter Sunday or thereabouts, exactly a quarter of a century since Ponce de León had given most of western North America the name Florida, he took some men ashore. The officials of the expedition now formally took possession of the land for the Spanish crown. In due course, they performed the serious but ludicrous business of reading out the Requirement and then set about making their first tentative explorations of their new surroundings.

In 1536, Cabeza de Vaca would tell Viceroy Antonio de Mendoza and the Mexican Audiencia that he remembered the trouble they had communicating with the first Florida Indians they encountered when they went to the men's village and explored other settlements where they were shown a few fields of cultivated corn. They also came across the pitiful spoils from a wrecked Spanish ship, a few rags and worthless pieces of cloth. But along with this debris, they also found some merchant's packing cases of a kind familiar from Spain, each carefully packed with a human cadaver. We can only assume that these dead men must have appeared to be Indians, because Juan Suárez, the religious leader of the expedition, ordered the soldiers to burn this macabre find, without showing the slightest vestige of intelligent curiosity in either the dead or their coffins. Had they been Catholic Spaniards, of course, he would almost certainly have had them buried.

"We also found samples of gold," Cabeza de Vaca told the Audiencia of Mexico. "By means of signs, we asked the Indians where they had come by that gold. Their gestures told us that it was far from there, in a province known as Apalache. In Apalache," they said, or so the Spaniards understood, "there was plenty of gold and great quantities of other valuable goods," such as silver and pearls.

Such news of gold was like a bullfighter's red cape swung gracefully before the most noble of Iberian bulls. The Spanish conquistadors got their heads down in blinkered determination to seek out Apalache, there to drive their horns deep into the promised treasure trove. But just as the bull finds only air beyond the smoothly handled cape, so too their dreams of instant wealth were a mere chimera, figments of their own imaginations and desires.

Without Miruelo to temper their lack of firsthand knowledge, the pilots had convinced themselves and their commander that Tampa Bay was a few leagues north. Narváez now became quite determined to find it.

Cabeza de Vaca then told the viceroy and the Audiencia that the promise of gold was enough to blind Narváez to the perils of his situation. The imprudent governor general gathered his officers around him on May 1 and stated his intention to split the expedition, sending the ships along the coast while he led 300 soldiers inland in search of golden Apalache.

Cabeza de Vaca, well tutored by Zumárraga, told the Audiencia that he had opposed Narváez's foolhardy decision. It would be better, he claimed that he had argued, to keep together until they found a safe haven for the ships. Only then should they risk an overland expedition. What, he had asked Narváez, would become of the soldiers should they fail to find their ships again? The foresight of that objection may strike us at first as sagacious, but it is worth remembering that he told the story with hindsight. There were perhaps those in Mexico City who knew differently, men and women who had remained on the boats. But they kept their own wise counsel and their mouths tight shut. There was nothing to be gained by showing up Cabeza de Vaca as a liar.

With pride ruling over reason, Narváez now ridiculed his treasurer, suggesting he was afraid to venture inland. Faced with such a slur on his honor, Cabeza de Vaca had little choice but to acquiesce in his commander's insanity and march to Apalache, or so the story goes.

Narváez ordered his ships to sail north in search of Tampa Bay. In time, the captain left in charge of that flotilla, a man called Carvallo, realized that Miruelo had been right after all. He ordered the boats to turn south and soon found the entrance to Tampa Bay. But by that time Narváez had led his army deep into northern Florida. The two parts of the expedition never saw each other again, until the four survivors met a few men from the ships in Guadalajara, in 1536.

Narváez's exact motives for marching inland are difficult to fathom. Did he and his men really believe they would find another Mexico in the province of Apalache? Was the decision to march inland made out of necessity by a leader who knew his men would soon go hungry unless they found a source of food? Was action the leader's response to the bickering amongst the pilots and the unrest amongst men who had suffered such ill luck on the seas of the Caribbean? Was Narváez, in his own way, emulating Cortés's legendary decision to scuttle his ships and so deny his men the possibility of retreat?

At first, as the Spaniards proceeded north through difficult terrain, the scattered groups of Indians avoided them, for no doubt they had seen slave ships anchor off their shores before. The expedition found little corn, and demoralizing hunger now brought doubt and uncertainty where once there had only been confidence and bravado.

After a fortnight of this depressing march, Cabeza de Vaca recalled as he gave evidence, they came to a river, which they found very difficult to cross. On the other side, 200 hostile Indians confronted them. The Spaniards attacked and took half a dozen prisoners, who then led them to a nearby settlement where they found a welcome surfeit of ripe corn, ready to be harvested. Further fruitless exploration occupied the Spaniards until the middle of June.

An Indian chief called Dulchanchellin listened to reports of their progress with growing interest. This new invasion seemed more serious than the many raids for slaves that had troubled his people intermittently for a generation. He was intrigued, and around the time of the summer solstice he set out with great pomp and circumstance to introduce himself to these unusual foreigners.

The Spaniards soon became convinced that Dulchanchellin was a neighboring lord hostile to Apalache. Cabeza de Vaca recalled that gifts were exchanged and a pact was made to combine their forces for an assault on that imaginary land of riches. The reality seems to have been that this was a ruse to lead the Spaniards into an ambush. When the time came, Dulchanchellin and his retinue vanished into the night. Narváez then took some Indian prisoners as guides, but these brave spies led their adversaries into "lands that were laborious to cross, but marvelous to see. There were great bushes with wonderfully tall trees, so many of which had fallen that we had to make great detours." They were walking into a trap and were soon attacked. In the skirmish, a Mexican nobleman was killed, an Aztec adventurer who had accompanied the expedition.

Discord among the Spaniards, abandoning the ships, the rebellious local leader allied with the Spaniard against a wealthy lord, the threat of Indian treachery, even the image of a road blocked by fallen trees, are all features of Cortés's account of the early days of the conquest of Mexico. It is as though Narváez were trying to force History to repeat herself, but instead of setting his deeds of valor in majestic Mexico and the civilizations of Mesoamerica, he had to be content with a pathetic sojourn through the swamps and sands of Florida.

WHEN THE EXPEDITION finally reached Apalache, they found a settlement of forty grass huts, a few cotton blankets of indifferent quality, and plenty of mortars for grinding corn. The deluded adventurers awoke from their dreams of wealth and focused on their plight. They were hungry, lost, and increasingly exposed to attack by hostile Indians.

Esteban formed part of a search party led by Castillo, Cabeza de Vaca, and Dorantes, which eventually found the shore. They hurried back to camp, where they found that Narváez and many others were now seriously ill and had been harassed by increasingly aggressive Indians.

It was now late in the summer of 1528, and with the expedition marooned somewhere on the shores of Apalachee Bay, in modern Alabama, they knew that without some drastic plan, they were certain to die. They discussed their options until all were agreed upon a course of action that seemed "impossible to every man": to build five boats in which to sail to Nuño de Guzmán's settlement at Pánuco in northeast Mexico. They "had no tools, no iron, no forge, no tar, no cables"; they had "nothing that was necessary" for their purpose. "Above all, they had nothing to eat" while they set about the task.

The Spanish conquistadors of the sixteenth century may have been cruel in their treatment of Indians, they may have been greedy for gold and precious gems, and they may have been violent and murderous when they quarreled among themselves. But they stood together in adversity with a spirit of determination and personal bravery that often made them masters of their world. Narváez's men were now united by that single-minded sense of purpose. Even Esteban knew that his life depended on being part of that team. His status as a slave was beginning to be blurred by a situation in which social hierarchy counted less than ability and strength.

Ten years later, another group of ill-fated Spaniards discovered the furnace made by Narváez's men with which they smelted any iron-mongery the expedition could spare. They melted down stirrups, crossbows, perhaps even daggers and swords. From that metal they made saws and axes to work the wood and nails to hold their ships together. Under the directions of a Greek boat builder called Teodoro, the carpenters set about their business, working from dawn until dusk. They made ropes and cables from their clothes and by plaiting the horses' manes, and they devised caulking from palm leaves.

Esteban and the other slaves set about helping the infantrymen to work on the ships, while the cavalry officers raided the surrounding country for food. The weaker men went along the shore, fishing and collecting shellfish. Every third day a horse was slaughtered to provide nourishing stews for the sick, and the horsehide was made into wineskins to carry water on their voyage. There is still a place in Apalachee Bay marked on maps as the Bay of Horses.

At the end of September, the five craft were ready, each designed to hold about fifty men, packed tightly. With only a few inches of free-board, they gingerly set out, paddling gently westward along the coast. Cabeza de Vaca later recalled that they believed they had already traveled 280 leagues, about 900 miles, along the Florida coast. In reality, Narváez and his men were barely 300 miles from Tampa Bay. Worse, they had come to believe that the whole gulf coast was no more than 1,000 miles long, whereas the true length is nearer 3,000 miles. As they set out in their flimsy craft, full of hope, they thought they might be no more than 100 miles from Nuño de

Guzmán's settlement at Pánuco. That catastrophic error killed almost all of them.

Then, as they sailed on their strange boats, they suffered terribly from hunger and thirst, and there were some who died from quenching their thirst with seawater. They were battered by storms and attacked by Indians when they tried to land. Narváez himself was badly wounded when he and his delegation were attacked trying to trade beads and other trinkets for food with a local chief. Cabeza de Vaca was wounded too, while providing cover for Narváez's evacuation. Dorantes and two other captains fought a valiant, determined rearguard action alongside Esteban that finally held off the Indians. It saved their lives.

Famished and parched, tossed by the rough seas, exhausted by the searing sun, they were again forced to risk begging food and water of a tribe inhabiting the land around Mobile Bay. Now, the Greek shipbuilder and an unnamed black slave jumped ship in desperation and abandoned the expedition to seek comfort among those Indians. Years later, these maroons were both reported dead by another Spanish expedition that failed to conquer Florida.

The five rickety craft pressed on, with the men losing faith, day by day, in their own certainty that Pánuco must be near at hand. They reached the Mississippi, pouring sweet water far out into the sea; they drank their fill. It was a welcome but temporary relief. They sailed onward, past Louisiana, toward Texas.

14

THE ISLE OF MISFORTUNE

1528–1529

THERE WAS A surprise awaiting the large group of Karankawa Indians who were sheltering from a violent storm on the night of November 4, 1528, huddled together, crouching around the hearths of their open-sided wigwams. As the Karankawa had always done, they had migrated toward the Texas coast during the fall in order to spend the winter there after the long hot summer in the river valleys of the mainland, eating fruits and seeds and hunting bison and other game. Now, they had come out to one of the barrier islands, probably Galveston, where plenty of fish and cattail roots would keep them well fed as the bitter "Blue Norther" winds blew the cold of the Midwest down into the Mexican Gulf.

Karankawa Indians and their ancestors had long lived along the coast of Texas; for generations they had been almost amphibious hunter-gatherers, masters of their salty winter world of land, marsh, and sea. As yet, they knew little of Europeans. One or two of them had, no doubt, seen a passing ship, nothing more than a strange if wondrous sight on the horizon. Others had heard rumors of raiding parties, perhaps, for Narváez and his cronies had sent their slavers to the north coast of the gulf from time to time. But those savage, barbarian missions had barely touched the lives of the isolated Karankawa, living in their hidden world behind the barrier islands.

With the passage of the centuries, these Indians would develop a ferocious reputation among European colonists. In 1687, the Karankawa massacred the survivors of a French expedition that established a camp at Matagorda Bay. By the time of the Texan War of Independence, in 1835, they were renowned as ruthless raiders and cannibals, described by many as giants. Although this picture of semi-human savagery was largely an insidious, colonialist myth, it is notable that the Karankawa struck almost everyone who left a record of meeting them as exceptionally tall. The men are constantly referred to by travelers and colonists as strong and well-built, able to fight off many assailants. And archaeologists have found enough old bones to confirm that they were indeed an exceptionally tall group, with skeletons and skulls that set them apart from neighboring Native American tribes.

As THE STORM abated and Aurora rose out of the olive-dark Mexican gulf, bathing the seaward beaches of their island in a golden glow, the Karankawa set about the daily business of finding enough to eat. The men inspected their fishing traps, set in the tranquil lagoon that lay between the island and the mainland. Some no doubt repaired any damage that had been caused by the storm. Others waded out into the water with their bows at the ready, poised to shoot the large black drum, sheepshead, and other fish that came to shelter in the calm, shallow waters of the lagoon.

The women busied themselves with finding fresh water and uprooting the tubers that the Karankawa ate for starch. Others gathered firewood thrown up onto the seaward shores by the storm. As they searched the beaches, looking out across the open sea, these women no doubt stopped to wonder at a strange sight out on the shining horizon. In the hazy distance something loomed and lifted on the glittering foam of the white horses, floating and flying, rising and sinking, coming nearer and nearer. Soon they realized that it was some kind of boat, much larger than the canoes that their men punted about the lagoons. It was approaching their island, and it seemed to be filled with men. Small children were dispatched to alert the elders.

Andrés Dorantes, Esteban, Castillo, and the rest of their crew were desperate to land after their terrifying night at sea. They had been battered by the storm, barely able to keep their barge afloat, bailing furiously as the wind whipped the waves into great crests from which the little craft crashed down, time and again, into the deep canyons that briefly sank between the walls of water. Now, with the relative calm of morning, there were prayers of thanks to the Lord for the miracle of their temporary salvation from the wrath of Poseidon. But that joy and elation were quickly tempered by hunger and the sight of smoke on the horizon, a sure sign that this otherwise hospitable land was populated by Indians.

As they strained at their oars, Dorantes ordered caution. But the sea was still too strong for their vessel and their fate remained beyond their control. They were quickly caught on the fast-rolling breakers of the flow tide and brought crashing up onto the beach, stranded on the sandy margin of an unknown land. Those with the strength to do so dragged the boat high up onto dry ground, away from the rabid sea. But most were weak and painfully thin, their ribs clearly visible against their taut skin, their thin torsos showing beneath the tatters of their grime-encrusted clothes. They looked like medieval and Renaissance images of death herself, vividly portrayed by artists as a lifelike skeleton, with dangling flaps of rotten flesh, devoured by worms and beset with flies.

Among these living pictures of death there were no doubt some who had been in Mexico, or at least heard rumors that the Aztecs practiced human sacrifice and cannibalism. Fear spread fast among the weakened Spaniards, but while there were those who dreaded any meeting with Indians, it was clear that they must rest a long while on this desolate strand before returning to the sea. They began to set up camp, there and then, on the beach.

There is no record of how these survivors and the Karankawa first came into contact with each other, but it seems likely that Dorantes would have turned to those with the strength to make a foray inland. If such a party was indeed sent, Alonso del Castillo is the most likely candidate for leading it. Although young, Castillo was Dorantes's most senior officer, and the two men had together led early exploratory missions sent out by Narváez. We also know that Esteban was strong and reliable and that he would later prove to be brave, perhaps fearless when communicating with potentially hostile Indians. He too seems a likely candidate. Perhaps the first "European" these Karankawa met was, in fact, an African.

The tall Karankawa, with their matted hair, their brutal tattoos, and their bodies pierced by sherds of cane skewered through their breasts and earlobes, were the epitome of the European image of American savagery. Yet, despite their strange ways, these gentle giants soon comforted Dorantes and his men. The Indians turned their heads, averting their gaze from these unknown creatures who had floated out of the sea, their ragged clothes flapping about them like the skin of a molting snake.

The Karankawa tried to converse with these outsiders in the short-winded accents of their foreign tongue, punctuating their peculiar, alien prose with melancholic sighs. Esteban and the Spaniards did their best to understand the Indians and to explain their own plight, and when their gestures of hand to mouth brought agitated action, they understood that the Indians were eager to help. Soon the Indians came with fish and roots to feed the washed-up adventurers.

Esteban no doubt eagerly devoured his share of that native generosity. He and some of his companions had, perhaps, already cast about the beach in search of sustenance and found little to appease their hunger. Above the beach, the desert dunes seemed nothing but grass and barren trees. The Karankawa were masters of a difficult terrain, rich in food as well as dangers, but for the foreigner, the shores of Texas are a bewildering world, a "maze of islands, lagoons, and salt marshes." The low flat ground, the thick vegetation of the marshlands, and the endless interplay of land and water were the domain of the Indians alone.

The warriors had come to investigate the alien arrivals, but now they saw there was no danger and brought their wives and children down to the beach. The Karankawa had come to gawk at the foreign freak show. They wondered at the flimsy craft in which their guests had survived the violent winter storm on the open seas. They gladly accepted the baubles, the almost valueless colored glass beads that Spaniards used as currency with Indians.

Then news came from the other end of the island. Another unusual canoe had appeared, also thrown up on the beach, and more aliens had landed. One of these strange creatures had climbed a tree and then followed one of the paths that had been cut through the long grass and bushes.

This man was Lope de Oviedo, the most able-bodied of Cabeza de Vaca's men. They had survived the storm, but without the strength or will to manage their oars, they had drifted aimlessly in the rowdy seas all day and into the next night. In the dead of darkness, the dread sound of breakers alerted Cabeza de Vaca and his pilot that land was near at hand. They cast a plumb line to a depth of three or four fathoms, and the few who had the strength rowed with all their pathetic might, turning the prow away from the shore for fear they be dashed on the rocks. But the sea caught them up and threw them down again with such a blow that the torpid, starving men stirred into action. They struggled with their boat to beat the ripping surges that dragged them down. They breasted the swollen, rolling white caps, looking for a firm foothold on the sandy bottom. Somehow, they kept themselves above the waves, and with renewed vigor they rowed with flailing strokes until they reached the land alive.

The crisp and golden light of dawn rose upon a sorry scene. The men were eating a meager meal of toasted maize and slowly recovered their strength. Presently, as morning broke and the long shadows shortened, Cabeza de Vaca sent Lope de Oviedo to see what he could see. He wandered off and climbed a tree and there he deluded himself that the tufts of the island grass and pockmarked dunes had been caused by the hooves of cattle. In his foolish excitement, he rushed to tell his companions that they had been blown by fortuitous winds to a land of Christian cowboys. He then rushed back to his vantage point by the tree in order to search for some path or road and found an Indian trail, which he followed for a mile or two along the island, until he came to some deserted shelters. The hopes and fears of this human flotsam were no doubt as variable as the weather. Instead of some Christian farmstead, Lope de Oviedo found a cooking pot and some food: some small creature that the Spaniards referred to as a dog and some fish.

The intrepid castaway made off with the Indians' dinner, but soon saw three Indians coming along the path who called out to him. He gestured for them to follow him. The panicked Spaniards now gathered on the beach for comfort, while the three Indians waited and watched for half an hour, until a cohort of fully armed archers arrived.

Shipwrecks reports that the Indians seemed like giants to the Spaniards, although it also explains that Cabeza de Vaca could not tell whether this was because the Indians really were large or, whether they seemed giants because the Spaniards were so terrified. It is a marvelous literary moment that perfectly evokes the helplessness of Cabeza de Vaca and his companions.

Shipwrecks tells us that the Spaniards had every excuse for shunning battle with these Indians, for only six men had the strength to stand. But Cabeza de Vaca was the hero of his book and he alone, he said, approached this terrifying army. Each side did its best to reassure the other of its good intentions and the Spaniards handed over some cheap rosaries and tiny brass hawk's bells as a sign of peace. Each of the Indians, *Shipwrecks* claims, handed a single arrow to Cabeza de Vaca in exchange, and they indicated that the following day they would bring the castaways something to eat.

As good as their word, the Karankawa appeared as golden dawn rose from the olive-dark sea and her honeyed fingers touched the white sands of the beach. They came with plenty of fish and cattail roots, as promised, and then they came back again, bringing more food and their women and children to see the straggling Spaniards. They went away with a wealth of gaudy trinkets that the Spaniards gave them. But as yet, the shrewd Karankawa left the two groups of Spaniards in ignorance of each other.

The Spanish conquistadors are nowadays rarely praised—in fact, they are deservedly maligned—but their characteristic perseverance and determination should be admired. With the bare sufficiency of food supplied them by the Karankawa, Cabeza de Vaca and his men set about launching their boat again, to continue their journey to the longed-for settlement at Pánuco, somewhere in the west.

They stripped down to their undershirts and carefully tied their bundles of belongings inside the boat. They dug it out of the sand and manhandled it down the beach and into the water. They rowed out bravely into the wild sea, but a great wave crashed against the bows and flooded the gunwales. The oarsmen lost their grip on the oars and they were suddenly turned topsy-turvy by another wave washing over them. The tremendous winter seas disgorged them again on the beach, but the waters took their toll, for that day they lost the boat and three men.

As the sun set beyond the mainland and the white-horses winked orange out of the olive sea, the Indians came down to the beach to find their guests broken in mind and body, more dead than alive. By those waters they all sat down and wept, Spaniard and Indian alike, at the evil turn of fate, so that the sound of their lament could be heard far and wide. Then the Indians escorted the bedraggled Spaniards to their camp. They had lit fires along the route against the cold and they half carried, half hurried the weary Spaniards to the comfort of a large shelter.

THE KARANKAWA NOW sent word to Dorantes and his men, who were down at the beach camp, making repairs to their boat and preparing for an early departure. Communication was always slow and uncertain with the Indians, but the message must have been clear enough. There were other survivors; apparently one of the other boats had also washed up on these godforsaken shores. That could only be good news, as two boats were more secure than one. In the morning, Esteban, Dorantes, and Castillo set out for the Indian camp with a band of men.

Their optimism at finding that some of their companions had survived soon turned to grim pragmatism. Cabeza de Vaca's men were fearfully weak and had nothing—no clothes, no trinkets to barter, no weapons. They were very evidently a burden and as the expedition treasurer gave an account of himself and his men, it became clear that they were a burden best left behind.

Cabeza de Vaca explained that his boat was lost. The state of his men was enough to explain their predicament.

What of Narváez, the Captain General? Dorantes asked.

Zumárraga had seen to it that the four survivors should portray Narváez not only as foolhardy and incompetent, but also as a cowardly leader worthy of the most vile contempt. Cabeza de Vaca later told the Mexican magistrates that a few days before, as the ferocious seas had given way to relative calm, he had managed to steer his boat toward the governor general's faster craft. The two vessels had then headed for the shore with all hands rowing hard, but Cabeza de Vaca's men had been weakened and their boat sat low in the water. They knew that they could not make the beach without help from the stronger, swifter vessel.

"Throw us a line," Cabeza de Vaca had asked of Narváez.

"Every man for himself," was the impious reply recorded for posterity.

What Cabeza de Vaca really told Dorantes, Esteban, and Castillo about Narváez's conduct when they met at Galveston remains uncertain. But for Cabeza de Vaca, this was a difficult point in his story as he gave evidence to the viceroy and the Audiencia.

"Together, we determined," Cabeza de Vaca testified, "that we would repair Dorantes's boat and that those of us who were strong enough would sail onward. The others should remain on the island until they had recovered. And when God saw fit that they might be able to join us in a Christian land; then they could follow us along the coast."

It was clear to everyone in the courtroom that Cabeza de Vaca, like Narváez, had decided to abandon his men. Narváez's final instruction had been "Every man for himself," and Cabeza de Vaca seems to have taken that order to heart.

Dorantes's responsibility was clearly to his own men. So although we will never know how many of his charges he decided it would be reasonable to maroon on the island in order to make room for Cabeza de Vaca and perhaps a few others, the likelihood is that they were few.

It was an uneasy moment for Cabeza de Vaca as he explained himself years later in Mexico, but he knew that discomfort would be fleeting, because the attempted launch of Dorantes's boat had soon ended in disaster. It had sunk quickly and they were all, once again, washed up on that ill-fated shore. Fate had determined that Cabeza de Vaca would not abandon his men. However, he was soon to be abandoned himself.

"We agreed to send four very strong swimmers along the coast,"

Cabeza de Vaca said, in the hope that they could reach Pánuco. But for the remainder, "given the state that we were in, we agreed that necessity was the mother of invention and that we would have to spend the winter on that island. Most of the men were without clothing and the weather was far too stormy for a long march." It would have been madness to try to "swim across the endless rivers and lagoons."

Then disease struck them brutally.

"After five or six days," the survivors explained, "the men began to die. After a short period of time, of the eighty men in total from both groups who landed on the island, only fifteen were left alive."

Memories of famine in Azemmour and his seaborne flight to southern Spain must have flooded Esteban's mind as he contemplated the awful scene before him. Once again he found himself among starving, sick men cast carelessly by fate upon a foreign shore where any faint, fleeting hope was dashed on the treacherous rocks of pestilence.

Among the dead were "five Christians who had remained at the camp on the beach and who reached such an extreme condition that they ate one another, one after the other, until there was only one of them left."

At some point, either in Mexico or in Spain, Cabeza de Vaca was asked for the names of these desperate sinners, whose memory was perhaps better forgotten than recorded for posterity. He gave their names as "Sierra, Diego López, Corral, Palacios, and Gonzalo Ruiz." Strangely, no one seems to have asked him which of these cannibal Christians was the last one left alive.

Tragedy was overtaking Narváez's men up and down the coast. Hunger, violence, and disease would soon account for all but Esteban, Cabeza de Vaca, Dorantes, and Castillo. But as the men lay dying in the Karankawa camp and down on the beaches, one of the priests who remained alive, a man from rugged and mountainous Asturias in the north of Spain, moved among those stricken soldiers of fortune, muttering prayers and giving the men their last rites as best he could. Castillo, whose father was a doctor, gave what comfort he could by drawing on half-remembered remedies vaguely learned at his father's side.

The Indians must have watched the curious incantations of these foreign shamans with a peculiar fascination as the priest bent low to hear the feeble confessions of the dying and blessed the cattail roots as the bread of heaven. And then the Indians also began to fall sick, and soon many of them died of "a stomach disease," presumably the typhoid fever or cholera which had struck the Spaniards, and against which the Indians had no immunity. The disease spread fast, until only half the Indians were left alive. The Asturian priest now went among his dying hosts, for it was his Christian duty to baptize these men before they died, so that their souls might go to heaven. Did he, one wonders, seem a sinister figure to the bereaved Karankawa, distraught at the loss of so many bretheren?

With fear and anger pounding in their hearts and vengeance coursing through their veins, the Karankawa rose up and came to seek revenge against the disastrous Spanish aliens who had brought them nothing but death and destruction. Their valiant warriors, so skilled as archers, now bore down on the fifteen traumatized, skeletal Spanish survivors who remained.

But suddenly an Indian chief spoke up, calling for the aggressors to desist, and with his words he stayed the Spaniards' execution. How can you think that these strangers are the cause of our sickness? Can you not see that they too have suffered the same deadly fate, so that only a handful of them are left alive among us? If they had brought this evil upon us, then they need not have perished as well.

With some such speech, a Karankawa elder saved the few of surviving Spaniards and Esteban from the Indian archers and their arrows.

"At this point," Cabeza de Vaca explained in the emotionless language of the courtroom, "I became separated from my companions."

Ravaged by disease, the Indians moved away from the island in small groups, taking the Spanish survivors with them. The Indian social system was devastated, and too few were alive and healthy enough to perform the sacred rituals for the dead. The Spanish expedition broke up, its hierarchy also smashed. Despite Oviedo's objections, the island was unquestionably a place of *malhado*, of "misfortune," "bad spirit," or "ill luck." The ragged groups of Indians and Spaniards retreated to the salt marshes on the mainland, where they could scratch a living collecting oysters.

These almost beaten conquistadors spent the winter dispersed about the waterlogged world of the Karankawa, desperately hungry and utterly helpless—because of their ignorance of that wilderness—except for the help of their Indian friends.

"The Indians who were holding me left the island and crossed over to the mainland in their canoes," Cabeza de Vaca reported. But "Alonso del Castillo and Andrés Dorantes, and the others who were still alive, were with some other Indians who spoke a different language and were from a different family." They were not to see each other again for five years.

THE MAGISTRATES AND the viceroy now turned to Dorantes and Castillo as their principal witnesses. They too had crossed the narrow strip of water to the mainland. But for all that the strangers and deadly diseases had disrupted the usual pattern of winters spent on *Malhado*, the Karankawa knew well enough that they would best survive the coldest months around the lagoons and barrier islands, where the fishing was good.

Dorantes and Castillo recalled how they had then set out again, no doubt with Indians as their guides, back across the lagoon, to another island. With them went the Asturian priest, two of Dorantes's cousins called Diego Dorantes and Pedro de Valdivieso, and, of course, Esteban.

This was almost a family group, the Dorantes "tribe," the master and his slave now struggling to survive side by side and on equal terms, the priest, the cousins, and the faithful friend, teenage Castillo, the doctor's son. Theirs were ties of blood and purpose, genetics and practicality. There was a bond between these men, an understanding, a subconscious adaptation to the Indian way of life that, in the course of things, would make Dorantes, Esteban, and Castillo invincible in the face of perils and adversity. In that winter of disaster and tragedy, in their desperation, they forged a sense of a common future.

15

TEXAS

1529–1533

IN THE SPRING, the Karankawa began preparing for their seasonal migration inland, and it became clear to Dorantes, Castillo, and the other Spaniards that it was time to move on. They had learned enough of the native language to know that these isolated tribes were going to move away from Galveston and the coast. But the Spaniards had little interest in going with them; their best hope was to continue west along the coast in search of Pánuco, keeping a watchful lookout for any passing ship that might rescue them.

But while the Spaniards were strangers to their new role as aliens, Esteban was no more an alien among the Karankawa than he had already been for many years among the Spaniards. And now, adversity had destroyed the social order; perhaps for the first time, Esteban began to experience the anxieties of freedom. He could travel inland with the Karankawa and forever escape the Spaniards' savage slavery. But why relinquish any lingering hope of seeing Africa again? Why forsake so soon the elevated status suddenly bestowed on him by their predicament? Could he turn his back on everything he had ever known and roam for all eternity among these alien nomads, living out some personal purgatory in order to avoid the gilded hell of the Spanish world? Had Andrés Dorantes promised him his freedom, perhaps?

The answer may be that Esteban now became an equal of the Spaniards. In fact, there is every reason to believe Esteban became the first among those equals. He was a man of wide experience and had seen many worlds and many different human ways. Esteban spoke the language of his mother and the many languages of his many masters. He had long been an outsider, an alien, a foreigner, a stranger. Experience had taught him the meaning of difference; he had learned that the battle between good and evil is fought everywhere and in every soul. He knew the meaning of hunger and of fear. The Spaniards no doubt turned to him for his experience of dealing with adversity.

Over that winter of 1528 to 1529, for the first time in the history of modern American soil, an African slave shed his bonds and became a man in the eyes of his former masters. Over that winter, the first African-American was born. Esteban must then have felt the first thrill of power.

This moment of epic personal transformation was succinctly but eloquently recorded in the testimony given to the Mexican magistrates by the survivors. They were asked to list the men who eventually set off from *Malhado* with Andrés Dorantes, to go along the coast in search of Pánuco: "Alonso del Castillo, Andrés Dorantes, Diego Dorantes, Valdivieso, Estrada, Tostado, Chaves, Gutiérrez, Asturiano the Priest, Diego de Huelva, Estebanico the Black, Benítez."

For the first time in the survivors' testimony and for the first time in the official record, Esteban is mentioned by name. For the first time in the history of America, an African is mentioned by name. In fact, of this list of largely long-forgotten people, Esteban was to become the most influential in his lifetime and the most important figure in history. Both in real life and in the writings of historians, the slave was to prove more important and more famous than his master. What is more, by a curious irony, it was the master's testimony that served to leave a record of the slave.

In the shaded cool of a Mexican courtroom, in the summer of

1536, Esteban was born into the virtual world of words and literature; he now became part of that ephemeral account of humanity that we know as history. There and then, Viceroy Antonio de Mendoza, the great conquistador Hernán Cortés, Archbishop Zumárraga, Cabeza de Vaca, and Esteban himself all bore witness as his name took its place alongside theirs in the pantheon of the unforgotten. His grave unknown, his flesh gone, his bones lost, Esteban now lives on in the modern mythology of the American dream, another heroic protagonist in some semi-documented legend about the origins of a stolen continent.

TOWARD THE END of March 1529, the Dorantes "tribe" crossed back to *Malhado*, where they were soon joined by six other survivors who had wintered on the mainland: Diego de Huelva, Benítez, Chaves, Gutiérrez, Estrada, and Tostado. On the island, they came across a man called Jéronimo de Alanís, a distinguished royal official, a notary or scribe, with twenty years' experience in the Caribbean. He had been appointed to his first position in Jamaica by Diego Columbus, brother of Christopher Columbus. But there was now no need for such a hired pen, an unnecessary official who could rubber-stamp a written record of their activities.

They also came across Lope de Oviedo, Cabeza de Vaca's intrepid scout. But both Alanís and Lope were gravely ill.

The Dorantes "tribe" abandoned them to their fate, leaving them behind on the island of *Malhado*. For all the blank pages Alanís filled in his lifetime with carefully scripted business contracts, official reports, and information about births, marriages, and deaths, he himself received only a cursory obituary: a note to the effect that "later, he died," made by Cabeza de Vaca in *Shipwrecks*.

Dorantes now bought his group passage across the Galveston lagoon with a sable cloak they had acquired from the Indians of Mississippi many months before. They then set out on foot along the mainland coast. It is far from clear whether they realized that Cabeza de Vaca was close at hand, also terribly ill. But if they did, then they took the decision to abandon him as well.

There is considerable debate over the precise route taken by the survivors after they left *Malhado,* but most scholars now believe they followed the coast as far as Corpus Christi Bay, where the Texan shore sweeps south towards Mexico. They must have stayed as close as they could to the open sea of the gulf so as not to lose their way and in order to remain in sight of any rescue ship that might be searching for them. This route was to take them along the mainland beaches as far as Matagorda Bay, through the area of modern Freeport. Their actual path has long been erased by the continuously shifting sands of the barrier islands and river mouths along the gulf coast.

As the men set out, the brambles were beginning to bear blackberries, and the sweet fruit must have been welcome after the long, hungry winter. But the weather was against them as they walked along the coast, keeping to the beaches. They reached a large river in flood, its waters rising quickly because of incessant rain. As they made rafts, they must have known that the crossing was fraught with danger, but they made it across. Then they walked nine exhausting miles farther along the sands, with no sign that the weather would improve, until they came upon another river in flood. It flowed so powerfully and with such fury that its sweet water reached far out into the sea, and again they hastily, hopefully made rafts.

The first raft made it safely across, with Esteban and Castillo aboard, all of the bedraggled crew paddling for their lives as they took the flimsy, improvised craft across the torrent. But the second raft was swept out to sea, for the men were too weak and tired to paddle; they had eaten nothing along all that coast but seaweed and some tiny crabs "which were all shell and no meat." Two of the men drowned, while two swam for shore. The others could only watch, wearily aghast, as one of their companions was washed out to sea on the wild current, a lone figure clinging to the makeshift raft. The castaway was carried so far adrift that Dorantes, Castillo, and Esteban told the Mexican Audiencia that the raft was three miles offshore. But, they said, when this man had drifted beyond the force of the river current, he made use of a strong landward breeze, struggling up onto the top of the raft and, using his body as a sail, he slowly regained the land.

Such ingenuity and caprices of fate were the arbiters of life and death among those alien trespassers on foreign shores.

As they mourned the loss of two more men and celebrated the seemingly miraculous survival of this third man of iron will and nerve, they came across another survivor, Francisco de León, who had spent the winter alone, eating oysters and seaweed. He gladly joined the Dorantes "tribe."

Stunned by trauma, loss, and joy, they continued along the coast. Ten miles or more farther on, near the mouth of yet another river, they found the abandoned wreck of one of the other boats they had built at the Bay of Horses, the one Alonso Enríquez had commanded. There was no sign of the passengers and crew. Among them had been Friar Juan Suárez, who, in his absence and without his ever finding out, had been officially appointed bishop of Florida by the Spanish authorities in Europe.

They inspected the wrecked barge for signs of other survivors and probably spent the night there, but they found nothing to give them hope. There was nothing for it but to continue their journey in the morning.

After a trek of fifteen or twenty miles, which must have taken them a day or more, they reached another large river, probably the Colorado. There, they came across a group of Indians, who immediately fled, leaving behind two wigwams. As they contemplated yet another precarious river crossing, a group of Indians appeared on the opposite bank of the stream. These Indians had already had dealings with the men from Enríquez's wrecked boat and were familiar with such strangers.

The Indians armed themselves and, thus protected, they brought the survivors across the river and took them to their encampment. There, despite the fact the Indians themselves had almost nothing to eat, they were hospitable and generously gave the Spaniards a meal of fish and a bed for the night.

In the morning, the party set off on the fourth day of continuous travel since leaving Galveston on April 1. Over the next four days, two men collapsed as their debilitated bodies broke down under the effort of their relentless progress westward. The others were too weak to help or even bury their lifeless companions. With each day, their sense of humanity and civilization that was such a source of aristocratic, European pride gave way to casual barbarity, savagery, and a necessary disregard for their fellows. Esteban perhaps recalled the execution of a young man in Jerez in 1522.

We can imagine that Mendoza and his magistrates were themselves weary of this endless tale of woe. The loss of so many minor lives was of no consequence to the imperial machine. No doubt Esteban, Dorantes, and Castillo well remembered where and when their companions had died, but details of those deaths are not recorded in the official account. The crown officials were more interested in useful, geographical facts than in the meaningless names of lost proletarians.

By the end of the seventh day, a total of 120 miles from *Malhado*, the few survivors reached a great expanse of water. "We believed," Dorantes and Castillo explained, "that we had reached the River of the Holy Spirit, because we came across a wide bay that was nearly three miles across its mouth. It had a spit of land on the Pánuco side about a mile long, which has very large white sand dunes that must be visible many miles out to sea." They were finally on the tip of the Matagorda Peninsula, looking out across Matagorda Bay.

They were making remarkable progress.

The nine remaining men rested in this bleak place for two days, attempting to regain strength, eating whatever they might scavenge. They did not report eating fish, although we have to assume that they did, for they must have learned basic survival skills from the Indians at Galveston. Even so, their bodies were beginning to swell with malnutrition because much of their diet was indigestible.

As they rested, they slowly repaired, as best they could, a damaged canoe that they had found near the shore so they might cross Cavallo Pass onto modern Matagorda Island, a long strip of land that is one of the most spectacular nature reserves along the Texas coast. Matagorda Island today is nothing like it was when Dorantes and Esteban passed that way. Late seventeenth-century maps show that this single long strip of land was once a labyrinthine archipelago: islands set in shallow lagoons, swamps, and marshy ground. The survivors now ferried each other across Cavallo Pass and carried on, as best they could, along the coast.

After thirty or forty miles, they reached a narrow pass where another, smaller lagoon opened into the sea. "It was no wider than a river," they recalled. "The following morning, we saw an Indian on the other shore, but although we called out to him, he would not come, but went away. In the afternoon he came back and brought a Christian with him, a man called Figueroa, one of the four men we sent the previous winter to see if they could reach Christian territories."

"And then," Dorantes continued, "the Indian and Figueroa crossed over to where we nine survivors were. There and then he told us that his three companions were dead."

Two had died from hunger, but the third had been killed by Indians, he explained.

Dorantes and the other survivors were alarmed.

As yet the Indians had been mostly welcoming, at least at first. But those friendly Indians they had encountered at Galveston were in fact peripheral to the main Karankawa culture, which was centered on Corpus Christi Bay. Now they were heading deep into the Karankawa heartland.

But Mendoza and the Mexican magistrates were more interested in the secondhand account of Figueroa's story. They wanted a report on the fate of Narváez himself.

"Figueroa told us," Dorantes and Castillo continued, "that he had come across a Spaniard called Esquivel." Esquivel was the last man left alive out of all those who had sailed on the other boats captained by Alonso Enríquez and Narváez himself. "All the others died of hunger," but he survived, according to Figueroa, by "eating the flesh of the men who died."

Enríquez and his men had come across Narváez, still on his boat, farther down the coast. Narváez had then ordered his men to go ashore so that the survivors from the two boats could travel along the coast together, while he followed in his barge, ready to ferry them across the lagoons, rivers, and passes where necessary.

At the Bay of the Holy Spirit, *Shipwrecks* reports, Narváez then revoked Enríquez's command and put one of his own captains in charge of the group that was on land. That night, according to both accounts, Narváez, now very ill, had remained resolutely aboard his barge, without food or water. He was perhaps terrified that his own men would leave him behind. He need not have worried, for fate had stayed her hand too long. In the dead of night she crept up on him in the form of a northern breeze. As the wind got up, the boat began to drag the stone that served as an anchor and gently floated out to sea, with the captain general, his page, and his pilot aboard. They were neither seen nor heard of again.

The men left behind on the shore now blundered deep into the waterlogged world of the Matagorda coast. They gathered wood, threw together some basic rafts, and crossed the lagoon, making their way inland. There, in a swampland that they did not understand, they became prisoners of their own ignorance and were trapped by misery and despair. In the forests and thickets of their watery jungle, Narváez's newly appointed captain lost control of his men. They set upon him. They killed him. They ate him.

"Esquivel butchered the last man to die and by eating him, he kept himself alive until the beginning of March, when he was rescued by an Indian," the official report would eventually read.

This was much worse than what had happened at Galveston.

In Mexico, the listeners were now spellbound. They wanted to know what had happened to Hernando de Esquivel, the Christian cannibal of Badajoz.

"About a month later, more or less," Dorantes and Castillo replied, "we learned from some Indians with whom he had been living that they themselves had killed him because he ran away from them."

WHEN ESTEBAN HEARD Esquivel's story of Christian cannibalism and remembered the young would-be cannibal of Jerez and thought about the men who had resorted to cannibalism at Galveston, he must have wondered whether his companions would choose to sacrifice him first, when the time came. Spaniards evidently surrendered easily to their savage instincts, he must have thought.

"We were a short time with Figueroa," Dorantes and Castillo recalled, "listening to that awful news, but the Indian did not want to leave him with us and forced him to leave."

This is another astonishing reversal of the usual histories of Spanish heroism during the conquest of America. Instead of the typical story of a how a few valiant conquistadors defeated hundreds if not thousands of Indians, Dorantes and Castillo told the viceroy that one Indian had been able to impose his will on ten Spaniards.

But the reality appears to have been more complicated than the story they told. Probably under pressure from the Mexican magistrates, Dorantes and Castillo admitted that only two of their party could swim and that these men—the Asturian priest and "a youth"—did in fact cross the lagoon with Figueroa and the Indian. "These two went," the survivors claimed, "with an idea that they would bring back some fish" that Figueroa and his Indian said they had back at their camp.

In the morning, the youth swam back across the pass, bringing a little fish for his former companions. The Indians, he explained, were leaving that day to collect some kind of plant, which they brewed into a drink. They insisted that the three Spaniards who could swim should go with them, but wanted nothing to do with the hapless men stranded on Matagorda Island.

The following day briefly brought some semblance of hope to the forlorn group of abandoned survivors. Two Indians were gathering berries on the far shore when they saw the Spaniards and hollered to them. These Indians then crossed the narrow strait and began to help themselves to the survivors' meager possessions, treating them "like people for whom they had little respect."

Years later, in Mexico, Dorantes and Castillo need not have mentioned this detail. But they had been well coached by Zumárraga and they knew Jesus Christ had shown that suffering was the way to salvation. Their own suffering would make the story of their eventual redemption more forceful and more Christian.

But Oviedo's account also betrays the compassion of these Indians. They may or may not have treated the Spaniards with "little respect" at first, but they later invited the strangers into their homes and shared their food with them for several days. And Oviedo lamented, "What a surfeit of hardship a man may suffer in his short life! Oh, unmentionable torments wreaked upon a human body! Such unbearable hunger for those so weak of flesh! What excessive calamities cut them to the quick!" The royal historian well understood the meaning of their nadir, the importance of their degradation and suffering to the telling of a good tale. All this was highly emotive.

Dorantes testified to the Mexican Audiencia that he was certain only God could have sent them patience enough to atone for their sins through such suffering. The message was clear. It may have been a high price to pay for the folly of their greed, and ambition, but their due penitence must be endured to achieve absolution. Their terrestrial perdition was a gift from God that might earn them an eternity in paradise. "And," Oviedo would add, "because they were Spanish noblemen, men of moral worth who had never known such poverty, their patience needed to be equal to their hardships and their suffering in order to survive."

The link was now complete between endurance and suffering on the one hand, and virtue and nobility on the other. Today, this is an unfamiliar image of the Spanish conquest of America. It seems to us a fitting role reversal, almost a haphazard revenge for the many atrocities of empire. But this ignominious suffering was the experience of thousands of Spaniards who set out for America in search of personal dreams, escaping from the poverty of their daily lives, only for those dreams to be shipwrecked on the rocks of brutal reality.

In the Archive of the Indies I have read the stories of hundreds of failed lives told by forgotten men and women whose dreams became disasters, but who believed that to survive such failure with stoical pride was in itself an act of heroism. As Oviedo warned, the foolish hopefuls thronging the port and riverbank at Seville should seek the advice of those who have returned home from America, disillusioned and broken. There are few who do return, however, he pointed out, "for while the journey is long, life is short and there are innumerable opportunities to lose it!"

THE KINDLY INDIANS who had taken in Esteban, Dorantes, and the others soon tired of caring for these survivors they had rescued from Matagorda Island. "As happens anywhere," Oviedo commented acidly, "when guests outstay their welcome and all the more so when

they are uninvited and have nothing to offer." Castillo, Huelva, Valdivieso, and two others were forced away from the respite so briefly offered by their overstretched hosts. Those five suffering men shuffled
on through the swamps in search of another Indian band said to be
living on the shores of a lagoon twenty miles away. Of these five, only
Castillo would survive.

Oviedo's *History* reports that Andrés and Diego Dorantes remained at the Indian camp, along with Esteban, "whom it seemed to
them was sufficient for what the Indians wanted of them, which was
to carry wood and fresh water and serve them as if they were slaves."
In Oviedo's account, Esteban remained a submissive slave, accepting
stoically his duty to do the work of all three men as a kind of rent for
their stay with the Indians. But that representation of the situation is
so difficult to believe that at least one English translation of this
passage appears to misunderstand the original Spanish, rendering
the sense as meaning all three men, the two Spaniards and Esteban
together, became slaves to the Indians.

It is, of course, entirely plausible that years later, in Mexico,
Dorantes claimed that Esteban had loyally continued to submit to
Spanish noblemen even in their most dire straits, willingly serving
both the Indians and his European masters. Such a report would
even have reflected well on Esteban in the eyes of the Mexican officials, giving them the impression that he was a strong, true, trust-
worthy slave.

But it is possible to deduce, beyond all reasonable doubt, what must
have really happened on those unforgiving shores, for Oviedo goes on
to report that after three or four days the Indians threw out "these
others."

Scholars have assumed that by "these others" Oviedo meant all
three men and that Esteban must have been expelled from the Indian
settlement alongside Andrés and Diego Dorantes, as though the role
of the slave, by some natural law or by divine default, was to always

remain at his master's side. But as we read on, it becomes clear that in reality, Esteban had remained alone among these poor but hospitable Indians, after they had thrown out Andrés and Diego Dorantes. The explanation is obvious: Esteban worked hard, whereas the Dorantes cousins did not.

Esteban, after all, had lived the hard life of a slave and had starved in Azemmour. When he was offered hope by the sheltering Karankawa, amid such foreboding, he quickly did his duty, helping his hosts however and wherever he saw the chance. He was a hungry man glad to find men who would share their life with him. Eagerly he became used to the diet of "snakes, lizards, field mice, grasshoppers, cicadas, frogs and all kinds of vermin" that were to be had. There were spiders, too, and ants' eggs and earthworms. There was a popular saying in sixteenth-century Spain that "there is no better sauce than hunger," but Andrés and Diego Dorantes were oblivious of its meaning. They failed to work; they paid the price.

So Oviedo may have written the truth when he reported that the Dorantes cousins thought Esteban's labor would be enough to satisfy their hosts. But it was Esteban's choice to work with the Indians. He toiled not because his Spanish masters ordered him to do so, but because hard work was obviously necessary. He could see that he needed the help of their hosts in order to survive in a land of strange plants and animals. He knew perfectly well that in a country he did not know, he must make an alliance with the natives. Again and again, the Spaniards reported that at first the Indians were generous and shared their paltry supplies of food. Again and again, this generosity waned until the Indians became hostile. And, as Oviedo remarked, a useless guest is an unwelcome burden. So, while the Indians soon threw out the lazy Spanish aristocrats who served no useful purpose, they were happy to have the willing, hardworking Esteban as their guest.

For Esteban there was perhaps a breath of freedom in that prison without walls which was to be his home for many months. But what kind of freedom can he have felt, slave to his environment, beholden to his hosts, trapped between an endless continent and an endless sea? Freedom and liberty are not always easy to define.

Andrés and Diego Dorantes now wandered through the bog-lands for several days, lost and filled with a sense of hopelessness. Another band of Indians robbed them of their few remaining clothes. Then they came across the decomposing bodies of two of Castillo's companions, men who had been with them only a few days before. They had little choice but to move on. Soon, they found another Karankawa clan and Andrés chose to stay with them while his cousin Diego continued as far as a lagoon, in search of Castillo. But, instead of Castillo, Diego came across the other Dorantes cousin, Valdivieso, who told him about a fight that had broken out between some Indians and Figueroa and the Asturian priest. Later, Andrés Dorantes saw the priest's clothes, a breviary, and a book of hours, the treasured possessions of a pious man. But when the priest had returned to try to retrieve these possessions, he was murdered by the Indians.

ESTEBAN, CASTILLO, CABEZA DE VACA, and Dorantes spent five years living among the coastal Karankawa. What did they do? What did they think? Why did they decide to move on after five years and not before? When they did move on, they did not go to the Spanish settlement at Pánuco. Why not? Why on earth did they travel in the opposite direction? Why, instead, did they end up in Sonora in northwest Mexico?

Oviedo and *Shipwrecks* offer a bewildering range of differing and often conflicting answers to these questions. What is more, not only do Oviedo and *Shipwrecks* tell different stories, but neither account is especially clear in itself. Time goes astray. Whole years are missing. The anecdotal detail of individual incidents intrudes on broad generalizations, but with little sense of purpose. Important events are described differently and in a different order, so that they float like castaways on a narrative sea, washed hither and thither without much continuity of context. The most important of these floating elements in the story is the claim made in both accounts that somehow the survivors mysteriously became medicine men, shamans with almost magical, God-granted power over the Indians of the interior. But Oviedo and *Shipwrecks* do not concur as to how, when, or why this came about.

In spite of these confusing and at times contradictory sources, it is possible to shed light on what happened next. For five years, between 1529 and 1534, the survivors scratched a living for themselves as best they could. Most of that time, they were isolated from one another, each man being forced to surrender himself to whichever Indian band would have him. They were no longer treated as guests, but were forced to work hard alongside their hosts, each man his own island, a diminishing archipelago of Spaniards in the strange sea of Karankawa culture. Diego Dorantes lived and labored until the summer of 1531, when he was killed as punishment for "passing from one house to the next," although neither *Shipwrecks* nor Oviedo makes it clear why that should be a crime. Valdivieso and Diego de Huelva met the same fate.

Dorantes, Castillo, and Cabeza de Vaca, urged on by Zumárraga, described these conditions as slavery.

"We were forced into slavery," Dorantes reported of the survivors' life among the coastal Indians. "They treated us more cruelly than any Moor could have done. We were forced to travel up and down that coast, barefoot and without clothing, in the burning summer sun. Our business was to carry loads of wood and drinking water and anything else the Indians wanted and we dragged their canoes through the swamps for them."

But *Shipwrecks* states that among the Indians of Texas, "the men never carry heavy loads," because that was the business of women and old people. "The women are very strong and hardworking, and from dawn until dusk they dig up roots and carry firewood and water back to the camp." There is some indication that the survivors were forced to do women's work, for all that it seems illogical and improbable to us that women and old people should have done the heavy work in Karankawa society.

At first glance, we might suggest this was done to degrade the survivors, but these Spanish intruders were not the only men who busied themselves with female chores. Cabeza de Vaca claimed that while he lived among the Karankawa, he had seen "men who were married to other men," although his description of these feminized men is confusingly contradictory. He writes that "they go about dressed like women and do women's work," but he also explains that they could carry especially heavy loads and were much better built than the other men. Unlike women, they were also adept with the bow and arrow.

This is perhaps the earliest description of a practice, once widespread in Native American culture and common among the Karankawa, of allowing and even encouraging some people to cross the boundary between the sexes. Such people are often referred to as berdaches, although the term "two-spirit people" is preferred by some writers.

Berdaches were mostly men, and their sexual preferences seem to have varied between tribes and perhaps between individuals. Needless to say, Europeans have almost always assumed that these people were homosexual.

Sixteenth-century Spaniards were used to homosexuality, which they had been taught was a mortal evil, punishable by execution and referred to as the "unnatural sin." Men whose "love would dare not speak its name" lived perilously and theirs was a necessarily clandestine sexuality. So, when the Spanish survivors were forced by the Indians to do women's work, the experience must have been utterly humiliating and demeaning.

From the little we know of berdaches, scholars have been able to piece together a picture of a way of life that extended across most of

North America. The exact nature of the individual male berdache and his role in society varied from tribe to tribe, but these men who lived in part as women were often very skilled at women's work, which brought them considerable material wealth. They were respected members of their tribe whose ambiguous gender brought them power, influence, and a kind of talismanic status.

Men seem to have chosen the beradache way in adolescence, when female spirits appeared to them in dreams. In many tribes they were especially associated with the goddess of the moon, and because they crossed the rigid barriers between the sexes, they were held in awe as powerful, magical figures. They were a privileged class who could mix one gender with another because they had two complementary spirits, male and female. On a practical level, they could excel equally at hunting or doing housework. They also went into battle and, as Cabeza de Vaca reported, were often even stronger and more physically powerful than other men. Berdaches, as a result, were feared and treated with great reverence, and they have been associated with shamanism.

If the four survivors were forced to carry out the kind of burdensome tasks usually reserved for Karankawa women, were they in fact forced into the role of berdaches?

Anthropologists believe that it was common in some societies for men to become berdaches as a way of compensating for a lack of women. But why would there have been few Karankawa women?

Dorantes provided the answer in his testimony to the Mexican magistrates. "During the four years I lived among these people," he claimed, "I saw them kill eleven or twelve children, burying them alive. And that just goes for the male children," he added, for, extraordinary as it may seem, "they do not leave a single female child alive."

In primitive cultures infanticide is often a response to famine. It is a kind of birth control. In such extreme circumstances, killing girls is more effective in the long term than killing boys, because fewer females mean fewer children in the future.

In a society where there are many more men than women, some men have to do women's work. This may be what happened to the Spanish survivors and no doubt to a good many Karankawa themselves.

BY THE SUMMER of 1530, the survivors had separate lives, each eking out an existence as best he could, having occasional passing contact with the others. As berdaches or otherwise, they lived in this limbo until Diego Dorantes, Valdivieso, and Diego de Huelva were murdered by their Indian hosts. That frightened Esteban, Dorantes, and Castillo, and somehow they managed to communicate with one another and conspired to escape. They arranged to meet farther down the coast.

In August 1530, in the steam-room heat of the Texan summer, Andrés Dorantes fled his Indian hosts, who had scattered into the bush, setting about their seasonal business of gathering fruit and preparing for the hunt. He carefully waited until he was out of sight and then he ran for his life. The Indians did not follow him. Perhaps they did not notice him leave; perhaps they did not think him worth the effort. He soon came on another group of hunter-gatherers who called themselves Iguaces and "gladly welcomed Dorantes because they had heard that the Christians were hardworking servants," or at least that is the explanation Oviedo recorded in his account.

Esteban and Castillo, perhaps cowed by the commotion caused by their companion's escape, bided their time. Then, as the cold nights of November closed in and the Indians began to migrate toward the coast and the barrier islands, Esteban collected his courage and he too fled into the wilderness. Terrified, he crossed a wide body of water and struck out across the unknown territory beyond. Soon, in the depths of that wilderness, he met Andrés Dorantes, who was by now terrified of the Iguaces and was determined to go back along the coast.

It was a brief encounter between the former master and slave because, despite Dorantes's warnings, Esteban preferred to join the Iguaces rather than continue with his former master. Any trace of the bond between master and slave seems to have been broken. In time, Esteban was joined among the Iguaces by Alonso del Castillo, the doctor's son from Salamanca.

Meanwhile, Dorantes managed as best he could, but he struggled, claiming that he was forced to scratch a living for himself, surrounded by a malicious and hostile tribal group known as the Mariames. He was clearly incapable of integrating with their world or ingratiating himself with them in any way. He said that he suffered in abject terror, perpetually fearful that the Indians would murder him. Most of the time, he told the Audiencia, he tried to avoid them. Oviedo wrote that whenever the Indians "came across poor Dorantes, they were very aggressive, and sometimes, in fact often, they rushed up to him and drew their bowstrings back behind their ears, aiming their arrows at his chest. Then they laughed and asked him if he had been afraid."

SHIPWRECKS RECORDS AN unlikely story about Cabeza de Vaca's adventures during his five-year isolation from the other survivors. It is a curious piece of testimony in which he claims to have become so disillusioned with the hard life on Galveston that he escaped and turned his hand to trading among the Indians of the coast and the interior. As a merchant, he explained, he was treated well and earned respect.

Cabeza de Vaca's own account in *Shipwrecks* is hardly convincing. He claimed that the Indians came to respect him to such an extent that his fame spread far and wide, that he was well treated, and that they gave him food. Why, one is impelled to ask, did he not use this preferential position to proceed in the direction of Pánuco and return to civilization? If, as he claims, he had the run of the coast down to the Bay of the Holy Spirit and traded with the Indians there, why did he not come across Dorantes, Castillo, and Esteban on one of these journeys?

These questions were apparently asked by someone in Mexico, or perhaps later in Spain. Cabeza de Vaca found himself compelled to offer an implausible answer. "The reason I waited for so long" to escape "was so that I could take another Christian with me who was still on the island, a man called Lope de Oviedo." Each year, he claimed, he went back to the island to try to persuade Lope to leave with him, but every time Lope put it off for the following year. As one modern editor of *Shipwrecks*, Juan Maura, points out, such self-sacrifice would, unquestionably, have made Cabeza de Vaca a saint. But, while Cabeza de Vaca was many things, not all of them bad, he was definitely not a saint. After all, he had tried to abandon all his men at *Malhado* as soon as he saw a chance of escaping on Dorantes's boat. To have waited so long for Lope would have been wholly out of character.

It would be all too easy to dismiss his story of becoming a merchant as a lie, a self-aggrandizing addition made up for the Spanish court, if it were not that the same story was recorded by Oviedo, but with one crucial difference: in Oviedo, Dorantes is the merchant, not Cabeza de Vaca. Did one man copy an untrue story from the other, or is the story itself actually founded in fact? Was there really a foreign merchant trading among the Karankawa in the 1530s?

Oviedo was, without doubt, confused as he gave an account of Dorantes's life as a merchant. He clearly had trouble making sense of his sources. So much so, that he manages to have Andrés Dorantes in two places at once in his account. This has led scholars to assume that Oviedo's account was wrong and that Cabeza de Vaca had been telling the truth. But there is a problem with this reading that has never been properly dealt with.

Oviedo was a conscientious historian who was scrupulous about his sources. He clearly realized that there was something wrong with the material he had in front of him. He was, we think, working with Andrés Dorantes's partially revised copy of the official report. It was a much-traveled manuscript, and he seems to have come across a passage of testimony that was out of its place in the proper order. This may have been a loose sheet of paper that had been carelessly misplaced in the document. It may have been a mistake made by the person who copied down the version he had in front of him. We are unlikely to ever know the answer to these questions, as the original is probably lost forever. But Oviedo tried to make sense of the confusion by using the material he had in front of him. He worked hard to make it fit the facts and the context. And, after struggling with the material, he had clearly concluded that the merchant had been called Dorantes, and to make that point as plain as possible he stated it on three separate occasions.

But although Oviedo was confused and his lack of clarity affected his account, it is possible to demonstrate that neither Andrés nor Diego Dorantes could possibly have been the merchant.

However, there is a third "Dorantes" who has been overlooked. Esteban was frequently referred to in the contemporary documents as "Esteban Dorantes," because it was common practice for a slave to be given his master's surname. So, perhaps Mendoza and the magistrates asked Esteban what had happened to him during this period and perhaps this is what he told them? Perhaps Oviedo's sense of hierarchy then led him to overlook Esteban completely, leaving him confused because he could not fit either one of the Dorantes cousins into the role of the merchant?

I have no idea of the true identity of the Spanish merchant. It seems to me to be equally unlikely that Esteban, Andrés Dorantes, Diego Dorantes, Cabeza de Vaca, or anyone else lived for five years as a renowned merchant, plying his trade in shells, flints, and animal pelts throughout the Karankawa nation. But history has always accepted Cabeza de Vaca's claim that he was that merchant because that story was published in *Shipwrecks* in 1542. It seems to me that Esteban has an equal claim to that role, for what it is worth. If whimsy requires us to have a merchant, then let the first "European" trading venture in the geographical area of the modern United States belong to Esteban. Let the origin myth be that the first Old World business-man in Texas was African!

THE NOTION THAT Esteban Dorantes became the merchant may be appealing in all its multiple ironies, but it may be more useful to consider how Castillo and Esteban spent their time while nobody became a merchant and Dorantes lived in fear among the Mariames.

Shipwrecks and Oviedo have little to say about Esteban and Castillo during their stay around Matagorda and Corpus Christi Bay. But from the moment Castillo came across Esteban living with the Iguaces, the two men seem to have stayed together, and they evidently formed a strong partnership.

We cannot be sure how they spent their days, but they seem to have established a comfortable rapport with the Indians. Dorantes apparently told Cabeza de Vaca that while he himself lived in daily fear for his life, Castillo and Esteban were reluctant to leave the shelter provided by the Igauces Indians. Presumably their experience was more felicitous than Dorantes's calamitous misery.

The Iguaces perhaps took Esteban and Castillo down to the bucolic banks of the nearby river, where there were plentiful groves of wild trees that bore a fruit strangely like walnuts. Here, Esteban was perhaps the first African to eat that quintessentially American nut, the pecan, and the stream is still known as the River Nueces, from the Spanish *nuez*, meaning walnut. It is still land where myriads of pecans grow.

They helped to stalk the deer that came down to the coast. They played their parts as their new companions spread out across the landscape, just inland from the where the deer were grazing. At high tide, they advanced together, carrying burning torches, setting alight the undergrowth, forcing their prey down over the dunes and cliffs and into the olive-green sea. Then they waited patiently for the tide to ebb, leaving the drowned beasts washed up on the beaches below.

Esteban and Castillo sucked and chewed on the succulent venison, the meat of kings, salted by the sea and barbecued over the sweet, smoky coals of mesquite wood or slowly stewed over a low fire. They gladly joined the Indians in their ritual dances to celebrate the successful hunt, giving thanks to their ancestors, gods, and spirits. There they smoked tobacco for the first time, and, with the full moon rising, in the shelter of a large wigwam, they watched as the men brewed up the leaves of the yaupon trees to make Carolina tea.

The leaves were carefully roasted in the bottom of a cooking pot. Water was poured on and brought to the boil, taken off the heat, and then boiled up again. Finally, the Mariames stirred the liquid with a dried-out squash, scooping it up and pouring it back, cooling it, until a yellow froth foamed on top.

With grave demeanor the men drank freely from this vessel, passing it from man to man around the circle. Then one of the Indians rose and began to move around the fire, bent double, almost on all fours, his face and body masked by animal skins. The men began to chant, their voices rising and falling with a melancholy cadence. Some drew out gourd rattles and set up a beat; others made a droning noise, drawing a stick back and forth across a corrugated piece of wood, like a washboard. A flute was whistled in time to the rhythm. All night the Indians danced and on into the next day and then the next, until, exhausted, they collapsed to languish in a torpor of calm, lying about their camp, resting.

THERE IS A long "rogue" passage in *Shipwrecks*, similar to the merchant myth, which contends that on Galveston, *Malhado*, the Indians "wanted us to become doctors, without testing us in any way, nor asking us for our qualifications." It seems likely that this flippant but heavy-handed irony was inspired by a decree issued by the Spanish crown in 1535, the year before the survivors testified at the Audiencia, which ordered Mendoza to prevent people from "practicing as a doctor, surgeon, or pharmacist, or describing themselves as holding a degree or being a doctor unless they have passed the relevant university examinations."

Shipwrecks describes the Karankawa of *Malhado* as animists who believed that rocks, trees, and other objects each had a spirit. As soon as one of the Indians fell sick, they would call for a doctor, who would blow on the patient and use his hands to expel the sickness. If this did not work, he would cut into the patient's skin where the pain was and then suck at the flesh all around the wound. He would then cauterize the wound and blow gently on the crusting scab.

The Indians, we are told, insisted that the survivors take on the role of medicine men. "The way in which we healed the sick," *Shipwrecks* explains, "was to cross ourselves, blow on the patient, recite the Lord's Prayer and a Hail Mary, pray as best we could to God, and then make the sign of the cross over the patient."

But it is impossible to believe that the survivors first became medicine men on *Malhado*, in 1528. Neither Oviedo nor *Shipwrecks* mentions shamanism again until all four had abandoned the Karankawa and the coast of Texas altogether, six years later, in 1534. If they had been working as revered medicine men during that time, then their experiences on the coast of Texas would have been quite different from the conditions of "slavery" and isolation that they in fact described. What is more, any such experiences as shamans would have inevitably been recorded in the two accounts. The most likely explanation is that the long description of shamanism on *Malhado* was added to the account of their experiences on the island after the manuscript of *Shipwrecks* had been written. Moreover, that description of shamanism itself was either made up especially for that purpose, or—as I think more likely—was moved from somewhere farther on in the original text, the place in the story which corresponded with the real moment when the survivors first began to experiment as shamans. But what was the point of this major manipulation of the story, and who was responsible?

A compelling explanation, and the most convincing, is that when the publishers acquired the manuscript from Cabeza de Vaca, they were unhappy with the story as he had told it. They were well aware that the real fascination *Shipwrecks* would hold for readers lay in the bizarre stories of shamanism, miracles, and extraordinary survival that were buzzing around the courtly world in Spain. The miraculous nature of their salvation was the source of the survivors' celebrity. The traveler's tale, a tall story that was ostensibly true, would sell books and make money. But they had a problem: that exciting material made up only a small fraction of the manuscript, and it came much too close to the end. There were eye-catching accounts of cannibalism, dissent, mutiny, and death, to be sure, but they needed a fuller, more titillating version of events if *Shipwrecks* was to sell, and they needed that version to be nearer the "top of the piece," as it were. The publishers arranged for the necessary changes to be made, but in an age when printing was still an infant art and publishing an inexperienced trade, the results were far from seamless. Moreover, the editors' task was complicated by the fact that a year before the publication of *Shipwrecks*, Cabeza de Vaca had left Spain for South America, so he was not on hand to do the rewriting.

So where and when did the survivors really first become medicine men?

There is no conclusive answer to this question, because Oviedo and *Shipwrecks* treat the whole business of the survivors' shamanism quite differently. But the evidence, such as it is, suggests that Castillo and Esteban began to work as doctors, or, more accurately, witch doctors, among the Iguaces, in 1533 and 1534. Meanwhile Andrés Dorantes was suffering his personal purgatory surrounded by the hostile Mariames, and Cabeza de Vaca was isolated near *Malhado*. According to this theory, *Shipwrecks* and Oviedo are silent about Castillo's and Esteban's becoming doctors at this stage in the story because Cabeza de Vaca and Dorantes preferred to suppress their rivals' success in favor of their own accounts of suffering and isolation. It may be that the rogue passage describing shamanism at *Malhado* was in fact based on the testimony provided by Castillo or Esteban to the Audiencia in about 1533 and 1534, but was moved back to 1528 in *Shipwrecks*.

My reasons for thinking that Castillo and Esteban began to work as shamans or doctors while they were among the Iguaces will be fully explained shortly, drawing on the accounts of healing rituals and miraculous cures in both Oviedo and *Shipwrecks*. For the moment, I confine myself to making one point: it does make sense that during this extended period, when Esteban and Castillo were isolated among the Iguaces, they should have had time to understand the complex Indian culture of their hosts. And while Castillo understood rudimentary European medical practice, the Indians' animism was as natural to an African like Esteban as it was to foreign to men like Dorantes and Cabeza de Vaca.

The apparent assertion that Native American and African religions and belief systems should be comparable or similar runs the risk, of course, of implying that Africans and Indians were primitive peoples who shared a base, uncivilized humanity. This is the language of colonialism that was used to justify imperial oppression and explain empire as the civilizing mission of sophisticated Europeans. But my argument here is simply that European Christianity in the sixteenth century was intellectually determined to close off beliefs and ideas associated with other religions. This was especially true of Spain, where the Inquisition embodied a paranoid fear of spiritual difference. By contrast, American and African religions readily incorporated the gods and rituals of others into their own systems of belief.

Even the great Islamic lords and kings of the sub-Saharan world recognized the power of pagan spirits.

Esteban was much better equipped than either Cabeza de Vaca or Dorantes to become a shaman. Of the Spanish survivors, only Castillo, with the open-mindedness of youth and some rudimentary medical knowledge, was well placed to help him.

The archives of the Inquisition in Spain's former colonies contain thousands of documents that demonstrate how easily Africans and Indians adopted each other's beliefs and spirituality in addition to Christianity. But such spiritual promiscuity baffled and frightened the Spanish authorities, which, ironically, is why the Inquisition created the documentary record, preserving precious evidence of African-Indian relations.

A century after the Narváez expedition was washed up on the shores of Texas, the Mexican Inquisition became especially concerned about a black witch doctor called Lucas Olola. The record of that case is merely one among many, but it is a good example of how Africans and Indians interacted. The events that had concerned the Spanish authorities took place near Pánuco, only two hundred miles south of Corpus Christi, where Esteban and Castillo spent 1533 and 1534.

The Guastesco Indians of Pánuco formed part of a continuous Indian culture, known to ethnographers as the Western Gulf Culture, and they often performed a ritual in which they carried a complex flower arrangement that formed an effigy of one of their gods as they danced to the beat of a drum. The Inquisition had heard that "blacks, mulattoes, and mestizos also dance" to that god, and "in particular a man called Lucas Olola," a black slave who cured the Indians of their illnesses by "burning incense and sucking."

One of the witnesses called by the Inquisitors said he had seen Lucas Olola "dressed in the costume that the Indians used for this dance. He pretended to levitate, appearing dead, and then fell to the ground. He then stayed unconscious for a long time, blowing foam out of his mouth. Suddenly, he raised himself with great fury and said that the spirit had come to him." In that state, the document records, he would "walk through walls, going in and out of the houses."

According to the Inquisitors, all the Guatescos were convinced that illness and death were caused by witch doctors. They now believed Olola's claims to be a witch doctor, and there were reports that many Indians had died of fear because they believed that he could do them harm. The officials complained that he went about behaving like a "divine and powerful lord" and, like all witch doctors, "he robbed the poor Indians of all their possessions and even their women." They found him so terrifying that if he fancied some girl or other, her relatives—her father, her brothers, and even her husband—would let Olola take his pleasure with her.

There are evident parallels between Olola's story and the survivors' accounts of their own shamanism among the Indians of Texas. It is also worth noting that one account of Marcos de Niza's later expedition stated that Esteban had fallen out "with the friars because he took along the women who were given to him, and collected turquoises and accumulated everything," whether we believe that account to be an approximation of the truth or merely salacious gossip.

———— ➤ ◆ ————

16

SHAMANISM

1533–1535

EVERY YEAR, IN THE late summer, the relentless rotation of the seasons brought together all the Karankawa groups and other Indian tribes in an area somewhere south of the Nueces River. They came to eat tunas, the red-purple fruit of the prickly pear cactus, a sprawling, olive-green plant that is ubiquitous in the deserts of the American Southwest. These plants were a rich source of sustenance in the harsh world of the Texas Indians.

"It was the best time of year for these people," Dorantes told the Mexican Audiencia. "For, although they have almost nothing else to eat other than these prickly pears and a few snails they search for, they fill their bellies day and night. That makes them very happy, because for the rest of the year they waste away from hunger."

The same, of course, must have been true for the survivors. But they had another important reason to look forward to the prickly pear season. The great gathering of tribes and nations throughout the tuna fields was a vibrant social occasion, a time to make alliances and find a mate. As the Indians socialized, the four survivors also had the chance to spend time with each other.

During the prickly pear season of 1533, Dorantes, Esteban, and Castillo began to plan their next move. The ripe fruit made a convenient food for travelers, so this was the right time of year for a journey.

Dorantes later claimed that Esteban and Castillo were reluctant to leave the Karankawa lands and people. That was no doubt because they had established a good if simple life for themselves as medicine men among the ancient Texan tribes, while their conservative companions lived perilously, disengaged from the world and people around them. But it is clear from both accounts that all three men eventually agreed it was best to move on. The fact that they had not once glimpsed a ship on the horizon in the gulf was perhaps decisive in persuading Esteban and Castillo to move away from the coast. With no hope of rescue, they knew that they would be entombed forever in the endless circularity of life on the coast. They yearned for action, for difference, for change. They wanted to be masters of their own futures.

Their first attempt at flight failed when a group of Indians quarreled over a woman. The different Karankawa bands exchanged punches and battered one another with sticks, until, enraged, they all went their separate ways and took the survivors with them, one hither, another thither, so that they were apart once more. The survivors endured the repetitious life for yet another year.

Throughout the prickly pear season in the summer of the following year, 1534, the survivors bided their time, waiting for an opportunity to escape. "Often," Dorantes remembered, "we were on the verge of leaving, and yet it seemed our sins were our undoing, for we were split up and each of us was isolated from the others." They then agreed that they must risk everything before the prickly pears ended. They decided on the next full moon as their trysting night, when they were to meet at an inland heath that they knew would be filled with ripe prickly pears. Their plans laid, they returned to their respective Indian families.

Each man struggled with his fears as the new September moon

waxed each night, slowly unveiling her face as she rose evening by evening. Their eager impatience grew, tempered by paranoid reticence. Andrés Dorantes, by his own account, was again the first to take the risk and make the break. When he arrived at the rendezvous, he found the heath crowded with Indians who were eating the prickly pears. He approached them fearfully, but they welcomed him warmly. They were, it turned out, mortal enemies of his former tormentors, the Mariames.

Then, Oviedo reports, "after Dorantes had been there for three or four days, the black man arrived, following in his footsteps, and also Alonso del Castillo (for they came together)."

It is an important moment in Oviedo's hierarchically minded *History*, a point at which the veil of his social prejudice briefly lifts to reveal the truth. Oviedo typically relegated Esteban to an almost invisible role and described him simply as *el negro*, barely ever mentioning him by name and frequently making no mention of him at all. It is therefore completely out of character that when describing the joint arrival of the noble, aristocratic Castillo and the lowly *negro*, he should refer to Esteban first and mention the Spaniard only as an evident afterthought.

The explanation must lie in the actual events of late September 1534. Dorantes, as he gave his testimony in Mexico, vividly remembered his first sight of the strong, powerful, reassuring figure of Esteban emerging from the undergrowth to greet him. Dorantes was overjoyed to see Esteban again. By contrast, he seems to have remembered the moment when Castillo appeared, lagging behind, as of secondary importance. In the Audiencia, Dorantes, perhaps unwittingly, revealed his true emotions of five years before, and the established social order was turned upside down in his account. Once again, history can be made to uncover the truth.

Someone—either Mendoza, a magistrate, or Oviedo himself—was so surprised by this strange narrative that he double-checked the information, and a note was added in parentheses to confirm that Castillo and Esteban had indeed "arrived together," but that Esteban was described first in the account.

Dorantes had discovered from these friendly Indians that another stranger was waiting for them at a place farther on along the trail, clearly marked by columns of smoke visible in the distance. They all knew that the other fugitive must be Cabeza de Vaca, and Esteban and Dorantes set out at once in the hope of making contact with him before nightfall. The Indians were reluctant to let them go, and so Castillo remained behind. He reassured the Indians that his companions would soon return, explaining that they had gone to find their friend and bring him back. It seems that Castillo had, in effect, become a hostage.

As dusk gave way to night, Esteban and Dorantes came upon an Indian who led them to Cabeza de Vaca.

Providence now treated these resilient men kindly. In the morning, the Indians struck their camp and decided to go in search of a late harvest of prickly pears that, fortunately, was close to where Castillo was playing hostage. By this good luck the four were brought together once again. It was no time for hesitation.

"And so," Oviedo recorded, "believing that it were better to do their duty as Christians and as noblemen, which each of them was, than to live like godless savages, they entrusted themselves to Our Lord" and fled. "And Jesus Christ in his infinite mercy guided them, showing them the paths they should follow, while God tamed the wild hearts of those indomitable savages."

And so, that same day, they escaped and rushed headlong, with no idea of where they were going, through a hitherto hostile land. But just as Athene watched over strong Odysseus, so too the Virgin Mary or some pagan goddess of the Southwest now watched over the four survivors as the waxing moon nightly withdrew her presence from the wine-dark sky.

Although the four men found few ripe fruits on the prickly pears, there were enough to keep them going. Fear now drove them onward, until, in the late afternoon, "it pleased the Mother of God," according to Oviedo, "that the very same day, at sunset, they came upon some Indians of the kind they had hoped for. They were very gentle and although they had heard something of the Christians, they knew nothing of how badly they had been treated." This, the acerbic royal historian pointed out, "was a very good thing from the point of view of that bunch of sinners."

Shipwrecks describes this felicitous change in the survivors' condition in greater detail. "As we traveled, terrified that the Indians would follow us, we noticed smoke and we headed for it. Just before nightfall, we saw an Indian, but as soon as he saw that we were heading toward him, he ran off without waiting for us.

"We sent the black man after him."

The Spaniards told their tale so that it reflected well on them, and as their superiors expected to hear it. Esteban remains, in the words on the page, an exemplary slave, subservient even under such a strain. By the perverted mores of slavery, even Esteban benefited from this lie. Yet the reality, in Texas in the fall of 1534, must have been otherwise. There was no subjugation among these wanderers, brothers now in a venture of survival. If Esteban went on ahead, it was because he agreed to do so.

And that tells us a lot about Esteban. It tells us that he was brave and that he was confident in his ability to communicate with the Indians. It also tells us something quite fundamental about his relationship with his former masters. It tells us, in fact, that in practice Esteban had become their leader, at least in the sense that he went first and blazed their trail. In this spirit of "nothing ventured, nothing gained," he led the way, more intrepid than his former masters, more robust in temperament.

Because he went first, ahead of the party, Esteban now controlled the Indians' understanding and image of the survivors. He was the first to meet the Indians, to talk with them, to tell them who he was and who the white men were. From this moment onward, Esteban became the agent of the survivors' constant movement, negotiating with the Indians, choosing the roads they would take, the byways they would explore, and the nations and tribes they would meet.

For the Indians, the four survivors were an astonishing sight. They were amazed by the single mysterious figure at the forefront who was burned jet black by the southern sun. It was stranger still that the dark-skinned harbinger of hope should be accompanied by rude, ruddy, golden-skinned white men with flowing blond beards. But Esteban controlled that potentially powerful language of pigment and identity with precise dexterity and the Indians soon came to worship all the survivors as "Children of the Sun."

So, Esteban alone followed the frightened Indian who had retreated from the four survivors. As soon as the Indian saw that the figure approaching him was alone, he stopped and waited, full of wonder.

"We are looking for the people making the smoke," Esteban explained as best he could.

"I will take you to some houses near here," the Indian replied.

All four survivors now followed this Indian guide and by the light of the setting sun, they saw the Indian encampment, the smoke rising out of the creeping darkness. As they came close, they found four Indians waiting for them beside the trail, ready to welcome them as gods. "In the language of the Mariames, we told them that we had come in search of them. They seemed happy to have us among them and took us to their dwellings."

These Indians were known as the Avavares, and with winter approaching and the prickly pear season now finished, the survivors decided to spend the cold weather among these friendly people.

During the bleak winter of 1534 to 1535 they suffered terrible hunger, worse even than they had experienced during the years they spent on the coast. The Avavares were constantly on the move, in search of whatever food they could find, but often there was nothing to be had and the children "swelled up like toads" with malnutrition. Many nights death visited the tribe, and at first light they woke to find another Indian dead from disease, hunger, and cold.

They existed in that timeless world of mourning and survival, until August came, bringing another prickly pear season and very welcome relief from the privations they had suffered.

That summer of 1535, as soon as it was practicable to do so, Esteban set out with Cabeza de Vaca to make contact with a band of friendly Indians camped nearby. After a long day of travel during which they covered twenty miles across rough country without a trail, they reached the Indian camp. Three days later, Castillo and Dorantes caught up with them and they all then set out with the Indians to eat some kind of bean or pea, the fruit of some unidentifiable tree that has been subject of much scholarly speculation.

The beans were not yet ripe, and the survivors were once again desperately hungry. They managed to trade some of the animal pelts that had kept them warm over the winter for a couple of creatures described as "dogs." They sat down to breakfast on the meat of these animals. Even proud Cabeza de Vaca, who had refused to eat horse meat on the shores of the Mexican Gulf, now feasted on the flesh of these quadrupeds of uncertain provenance.

They then continued their journey. All that day they traveled across a desert empty of people and sustenance. Suddenly, they were overtaken by a summer storm and in the deluge they lost their way and were forced to take shelter in the dense bush. They dug out a pit oven and filled it with pads from the prickly pear cactus, which they roasted overnight so that their breakfast would be ready at dawn.

The following day they came across a large Indian encampment of forty or fifty wigwams or tepees. They had crossed an important cultural frontier, leaving behind the Texas Karankawa and entering the Mexican world of the Coahuiltecans.

"And it was there," Oviedo recorded, "that the Indians first began to fear these few Christians, holding them in great esteem and showing reverence for them. The Indians came up to them and began to rub themselves, gesturing to the Christians to rub and stroke them. They brought some sick people to be cured, which the Christians did," Oviedo remarked, perhaps with considerable skepticism, "for all that they were more used to hardships than performing miracles."

IT IS EVIDENT from Oviedo's account that the letter Dorantes posted in Havana, in 1539, first described the survivors' work as medicine men at this point in his story. Dorantes's first experience of the survivors' shamanism, it seems, was in the summer of 1535, when they initially made contact with the Coahuiltecans.

The same episode is also reported in *Shipwrecks*, although quite differently.

Because the editors of *Shipwrecks* had introduced a fictional first account of the survivors' supernatural powers into the description of *Malhado*, they had left an unbridgeable gap between that fictional episode and the real first experience of shamanism, whichever of the four initiated it and wherever it had really taken place, whether among the Iguaces, or here among the Coahuiltecans. The editors then tried to fill that gap with a more elaborate account of shamanism, which was based on the same real events described by Oviedo as taking place in 1535 among the Chuihaltecans, but which they now set among the Avavares, in 1534, so that the episode took place earlier in the book.

The Avavares Indians, according to the rewritten *Shipwrecks*, took the survivors to their camp, where Esteban and Dorantes stayed with one Indian doctor and Castillo and Cabeza de Vaca with another:

The very same night that we arrived, some Indians came to Castillo and they explained that they were suffering from terrible headaches. They asked him to cure them and as soon as he made the sign of the cross over them and prayed for them, they claimed that they were cured. Then they went off to their dwellings and came back bringing plenty of tunas and some venison and there was so much that we had no idea what to do with all that meat.

Shipwrecks tells, in due course, that many Indians arrived one morning, bringing five comrades "who were paralyzed and very sick." They came in search of "Castillo, so that he might cure them, and each of them offered him their bows and arrows. Castillo welcomed them and then at sunset he made the sign of the cross over them and prayed to Our Lord God." All four survivors joined him in these prayers, hoping for a miracle, beseeching God to bless the Indians with good health, for they believed it to be the only way of persuading those people to help them escape "such a miserable existence."

In the morning, the sick men, even those who were paralyzed, awoke fully recovered, astounding the Indians.

The survivors now moved on and joined another band who were eating prickly pears, and the Indians now flocked from far and wide in search of Castillo.

In *Shipwrecks*, Castillo at first seems, without doubt, the Indians' preferred medicine man. This is not surprising, as his healing powers seem quite extraordinary. It was a relatively mundane achievement to cure the severe headaches suffered by the first group of patients, but his fame spread quickly when the paralyzed people began to walk. In *Shipwrecks*, whether we believe it or not, we read about an astounding miracle of biblical significance.

But then in *Shipwrecks* we are told that these people had heard about the marvelous cures that the survivors performed while trading with the coastal bands of Karankawa—an account which supports the idea that Esteban and Castillo had first worked as doctors in 1533 and 1534, among the Iguaces, who were a Karankawa clan.

That argument seems, at first, to be further supported by the fact that Castillo is central to the account in *Shipwrecks* of these first miraculous cures among the Avavares. Yet the tone then suddenly changes. *Shipwrecks* now tells us that because word had spread far and wide about the survivors' miraculous cures, a delegation of Susola Indians arrived in search of the famous shaman, Castillo. They begged him to go with them to help others who were very sick, including one man who was on the verge of death.

"Castillo," Cabeza de Vaca now claimed, "lacked confidence as a doctor, especially when faced with a terribly risky and dangerous case. He was convinced that God would count his sins against him and that this would sometimes prevent his cures from working."

On the previous page, Castillo had worked miracles like those in the New Testament: paralyzed men had gotten up and walked. Now, a sentence or two later, he is robbed of that role and instead described as a shy, ineffectual shaman—and as a terrible sinner guilty of nameless crimes.

Our sense of what really happened begins to fade as the text becomes uncertain. According to *Shipwrecks*, the Susola Indians now turned to Cabeza de Vaca, because they remembered that he had performed successful cures a year or two earlier, during the pecan harvest. This is palpably nonsense: *Shipwrecks* gives a rather full description of the pecan harvest in which there is absolutely no mention of shamanism, of miracle cures, or of medicine of any kind. Even more damning, with regard to Cabeza de Vaca's claim, is the fact that in Dorantes's account, as transmitted by Oviedo, these first major episodes of shamanism began later, among the Coahuiltecans, who had never gone anywhere near the pecan harvest. But Cabeza de Vaca somehow had to explain to his readers why in the next episode he was about to become the main medicine man, or, rather, appear to become the main medicine man. I say "appear" because a careful reading of Cabeza de Vaca's account reveals that he did not actually record himself as being responsible for the miraculous cure at all. Because of Castillo's timid reluctance, "I had to go with the Indians," Cabeza de Vaca boasted. "And Dorantes and Esteban came with **me**. As **I** approached the Indian camp, **I** saw the patient **we** had come to cure.' Note that "I," Cabeza de Vaca, gives way to "we," Dorantes and Esteban, when it comes to doing the healing.

They were sure the man was already lost, "because he was surrounded by people and his wigwam had been dismantled, which is a sign that the owner is dead. And so, when **I** arrived, **I** found the Indian with his eyes rolled back and without a pulse. It seemed to **me** that he showed all the signs of being dead and Dorantes agreed.

"**I** removed a grass mat that was covering him and, as best **I** could, **I** prayed to Our Lord for the good health of the man and of all the others who were sick."

But while "I," Cabeza de Vaca, did all this, in the next sentence we read that "the man had the sign of the cross made over him and was blown on many times." Cabeza de Vaca makes no attempt to say that "I" made the sign of the cross, or that "I" blew on the patient. For some reason he tries to imply that he was the medicine man by suppressing the protagonist's real identity, rather than simply making a false claim. But, because of that curious honesty, he in fact implies that it must have been someone else who did the cure by shifting from "I" to the impersonal passive voice at the crucial moment. There is little question that had Cabeza de Vaca made the sign of the cross and done the blowing, then he would have said so loudly.

So, if Cabeza de Vaca did not make the sign of the cross over this dead Indian, did not blow on the patient, and therefore did not perform the cure, who did? *Shipwrecks* states clearly that Dorantes had by then not effected any cures—a statement which may or may not have been true but which is all we have to go on. Therefore, by a process of elimination, the only possible answer is that on this occasion Esteban was the doctor.

The three men now returned to Castillo, and that night some of the Indians who had accompanied them on their visit to the Susola returned, bringing news that they had seen the dead man up and about, looking well. They said he was walking, eating, and talking with them, and was fully recovered.

This struck awe and terror into Indians across all that land, and they now spoke of nothing else. By sleight of hand, Cabeza de Vaca gave his readers the impression that he had raised a man from the dead. But he did so without actually claiming directly to have either personally performed the healing rituals over the dead man or to have seen the results for himself. *Shipwrecks* appears to offer something of the truth, and it seems that in fact Esteban may have tried to perform the miracle, but also that there was no evidence a miracle had even taken place.

Between these two episodes of miraculous, biblical healing—between Castillo's cure of the paralyzed patients and someone's raising a man from the dead—the publishers of *Shipwrecks* offered their customers another marvelous episode, seemingly more fiction than fact.

The episode is described as happening to Cabeza de Vaca, but in reality it might have been experienced by any one of the four survivors, or it may never have happened at all. One of them became separated from the others while searching for food in thick scrubland and wandered lost for five days in the winter cold, without shelter. However, in the midst of the Texan bush, this solitary survivor came across a burning tree and was able to warm himself by this miraculous, heaven-sent fire. It saved his life.

It is not surprising, perhaps, that most scholars have interpreted this biblical "burning bush" as a figment of Cabeza de Vaca's imagination. Today, the story seems outrageous, but to Charles V, his Spanish courtiers, and the readers of *Shipwrecks*, it may simply have reinforced their impression of America as a New World filled with wonders and marvels. Interestingly, those presumably gullible Spaniards may have been right.

John Sibley was a physician from Louisiana who explored Texas at the beginning of the nineteenth century, when Texas was still part of the Spanish Empire. In his explorations, he came across the Karankawa, living in the bay of Saint Bernard on a strip of land ten miles long and five miles wide. On one side of their island home, there was, he recorded, "a high bluff, or mountain of coal, which has been on fire for many years, affording always a light at night, and a strong thick smoke by day, by which vessels are sometimes deceived and lost." He added, "From this burning coal there is emitted a gummy substance the Spaniards call cheta, which is thrown on the shore by the surf, and collected by them in considerable quantities, which they are fond of chewing; it has the appearance and consistency of pitch, and a strong, aromatic, and not disagreeable smell."

Below the surface of Texas is a reservoir of fossil fuels, ready to spontaneously combust at any time. So perhaps one of the survivors did indeed come across a burning tree somewhere in the thick scrub.

But the publishers of *Shipwrecks* had an urgent need to include wonders—a need which was not satisfied by burning bushes, cured paralytics, and resurrected cadavers. There was more. The Indians of the west Texas hinterland, we are told, believed in a creature called *Mala Cosa*, "Evil Thing." "He was small of stature, wore a beard, and they could not see his face clearly." He terrorized them, brandishing a burning torch, entering their homes, and stealing whatever took his fancy.

"He slashed three long cuts into a man's side with a sharp flint" and thrust his hand inside, carefully drawing out the entrails. He would then "cut off a small slice of intestine and throw it into the coals of the fire." This strange ritual continued, with Evil Thing engaging in further, equally peculiar practices. Sometimes he appeared dressed as a woman; at other times he dressed as a man. He never ate the food the Indians offered him. He could pick up a wigwam and throw it high in the air——yet despite all the violence he did to the Indians, afterward he would heal their wounds. He was a character seemingly straight out of the spirit world, and according to *Shipwrecks*, the Spaniards at first believed that Evil Thing was no more than a myth. But then they were shown victims who had scars from the monster's vicious attacks, and they began to believe the story.

The whole description might be a metaphor for the schizophrenic Spanish policy toward the Indians, with the brutal conquistadors on the one hand and the gentle missionaries and priests on the other. The survivors reckoned that Evil Thing had visited the Indians about fifteen or sixteen years before, leading some scholars to link this myth with the first Spanish raids and incursions into the area north of Pánuco and along the coast of the Mexican gulf.

Whatever the true origins of the myth of Evil Thing, the description of his bizarre behavior completes this series of wondrous episodes in which reality seems to give way to myth and fact is replaced by marvels. For page after page, *Shipwrecks* flirts with fiction, presenting the survivors' experiences between the fall of 1534 and the prickly pear season of 1535 as steeped in divine miracles and supernatural occurrences.

With the marvelous, wonderful episodes complete, *Shipwrecks* now falls into line with the history that was recorded by Oviedo. In contrast to the preceding picture of plentiful food supplies given to the survivors in gratitude for their miraculous cures, *Shipwrecks* soon describes the reality of mundane drudgery and the hard winter they spent among the Avavares. "We were always treated well, but we had to dig for the food we ate and carry our own loads of drinking water and firewood." They suffered even more than they had among the Karankawa. They spent their days naked, like their hosts, and at night they covered themselves with deerskins. The Avavares, we are told, "had neither corn, nor acorns, nor pecan nuts." It is the same picture of the Avavares that Oviedo paints for us——the true picture, one feels, and a picture from which miracles were conspicuously absent.

17

THE RIO GRANDE

1535

HAVING SPENT A terrible year with the Avavares Indians, the survivors moved on in August 1535, as the prickly pears began to ripen. Esteban and Cabeza de Vaca set out ahead, and the others followed three days later. The four men joined a band of friendly Indians, many of whom were very weak and sickly, with distended stomachs. There is no mention of any attempt to heal these ailing people, nor any indication that they asked to be cured. These Indians seem to have had no idea that the four were medicine men—confirmation, perhaps, that the account in *Shipwrecks* of their shamanism among the Avavares was untrue. The survivors bought some dogs from these Indians and cooked them, a welcome meal after the long winter of hunger among the Avavares.

Oviedo's *History* and *Shipwrecks* tell an almost identical story now, but *Shipwrecks* adds considerable detail that for once is entirely plausible. Leaving the sick Indians behind, they followed a trail that led into an area of thick scrub. It rained heavily and they were forced to shelter overnight. The following morning, as soon as the survivors were out of the bush, they found some wigwams and saw two women and some children who ran off, terrified, to call a band of Indians who were gathering food nearby.

These Indians arrived, but they stopped and watched the four strangers from the safety of some nearby trees. Eventually, Esteban managed to communicate with them, and the terrified Indians offered to take the survivors to their camp. That night, the strangers reached the camp, a large place with fifty dwellings. The Indians were frightened at first, but as they became more confident they came up to the strangers and touched them, stroking their faces and bodies. Oviedo tells us that it was here that the Indians "first began to fear these few Christians, holding them in great esteem and showing reverence for them."

"When morning broke, they brought us those people amongst them who were sick and asked us to bless them," Cabeza de Vaca recalled.

The Indians gave the survivors "whatever they had," but there was little to eat—just a handful of green prickly pears and some baked cactus pads. Even so, this was better than they had fared among the Avavares, and they rested for a fortnight with these Indians, recovering their strength. "The Indians treated them very well, willingly giving them whatever they had," Oviedo told his readers, before reminding them that "this was quite different from the treatment they had encountered before. For, until then, all the Indians they had seen or had dealings with had used them cruelly." But, slowly, as the prickly pears began to ripen, the four foreigners gathered their strength and readied themselves for their journey onward into the unknown.

Another group of Indians appeared, curious to meet the alien shamans, and Esteban and the other survivors decided to continue on with them. Their hosts were distraught and raised a great clamor of wailing and weeping. But nothing would change the four men's minds; they were determined to look for something better, to search for whatever the future might hold.

Their new companions led them to their nearby camp, where they formally welcomed the mysterious strangers. There was plenty of food: ripe prickly pears, and venison hunted by the Indians. The survivors responded to this diplomacy by healing the sick and blessing the tribe. Already, their fortunes were changing.

As they enjoyed this relative luxury, two emissaries arrived from a neighboring tribe. They were women, unthreatening and perhaps expendable ambassadors—a sensible, cautious choice as go-betweens. The briefest of conferences with these envoys was enough counsel for Esteban and the others to decide to make a long journey across inhospitable country, without roads or trails, to the people the women represented. The survivors now had a sense of urgency, and they prepared to set off. In their haste and confusion, the godlike strangers must have misunderstood the situation. Although the Indians urged them to stay the night so that the women might rest before guiding them along the right roads across the bewildering terrain, they instead set out immediately, and without their guides. That curious decision is not explained in either account, but is surely evidence of some unspoken strain or stress experienced by the four at this time.

Without the help of the Indians, they were soon lost, uncertain of where they were or where they were going. They stopped on the bank of a stream to ponder their predicament. Fortunately for them, as they rested there, paralyzed by indecision, they saw the women emissaries emerge from the scrubby desert accompanied by female representatives of their former hosts. The women had spent all afternoon looking for their foolish charges and they now roundly scolded the four men for wasting so much time and effort.

The following morning, the unusual party of travelers continued quickly, and that day they covered "eight or nine long leagues," thirty miles or so. In the evening, as the sun set hazily into the heat of the horizon, they saw the last golden rays reflected back at them from a wide river, "as wide as the Guadalquivir at Seville."

From the maelstrom of confusing ideas and self-confounding theories about the precise route taken by the survivors from Corpus

Christi to Culiacán, there emerges a rare sense of scholarly certainty that this river was the Rio Grande. On the other hand, there remains considerable uncertainty about how far up the Rio Grande they may have been. None of the theories seems very satisfactory, largely because there is so little concrete information about geography in either Oviedo's *History* or *Shipwrecks*. From Oviedo we can work out that the survivors themselves calculated they had walked about thirty leagues, or 100 miles, after abandoning the Avavares. We can also say it seems likely that during that journey they traveled in a more or less straight line. Beyond this we can say very little; we cannot be sure where they started, or in what direction they were headed.

We would do well to remember that maps and compasses, mathematically computed directions of travel, precisely measured distances, and carefully calculated rates of progress are western cultural concepts. In sixteenth-century Mexico and Spain, such scientific information was needed by cartographers, courtiers, and their king so that conquests, exploration, and claims to land could be marked on charts and given legal status. But for the four survivors now living as Indians, journeys were mapped mentally according to sources of food and water, the hostility of the people, and the number of days it would take to cross a particular terrain.

Somewhere between modern Laredo and the mouth of the Rio Grande, the four men now waded across that great river, left the territory of the modern United States, and entered modern Mexico. Ahead they saw a hundred tepees, clustered together in the wide expanse of arid semi-barren plain. As the party of shamans and their entourage of women arrived, a shout rose up from the people of this settlement, and the exuberance of that incomprehensible multitude worried the survivors. We may assume that Esteban once again steeled himself to lead the way, showing the Spaniards by example that the terrifying cries and screams and the dread rhythms of the dancers were sounds of welcome, not a war cry.

The Indian priests approached with gifts: rattles made of huge dried squashes filled with tiny pebbles. The gourds were brought downriver on the flood tides from far away in northern Texas and New Mexico, but these people of the plains believed that they fell from the sky like rain, presents sent by a friendly deity. They valued the rattles as divine treasure, displaying a mysterious, primitive human ability to revere something utterly ordinary—a primal urge for the impractical that the European world applied to gold.

These rattles would come to symbolize the spiritual power and mysterious provenance of the four foreigners for the remainder of their journey. These potent gourds were like western scepters, signs of shamanistic suzerainty construed from the vegetable flotsam of the Rio Grande. Later, they may have been the catalyst to Esteban's murder.

Their awestruck Indian hosts were as much thrown into frightened uncertainty and confusion as were the survivors. The fearful Indians crowded around them, pressing up against them, with those at the front reluctant and those in the rear pushing forward. The four were caught up and carried, like modern sports heroes, above the heads of the crowd.

All that night, the Indians danced and celebrated, but, according to *Shipwrecks*, the survivors resolutely remained aloof. The women from the previous tribe who had accompanied them across the desert were thanked with gifts of bows and arrows, and soon left. As day broke, the Indians abandoned their fiesta and came to feel the healing power of the Christians; they brought their sick to be cured and their children to be blessed.

The four men were now taken to another large camp that was nearby, and again, like touring rock stars, they were mobbed by adoring fans. Food was brought to them—the usual prickly pears and deer meat, an offering to the gods. This time the four men joined the night dance and began to cure the sick and bless the healthy. But their celebrity made for a relentless schedule. In the morning they moved again, to yet another Indian group.

Here, the survivors sensed a seemingly sinister intention in the behavior of their hosts. A crowd of men and women traveled the six leagues, nearly twenty miles, to the next encampment. When this entourage arrived, the four western gods were well received, although for three hours they were trapped by a pawing crowd. But almost as soon as they arrived, their Indian companions began to rob these enthusiastic new hosts, looting everything. When the survivors complained the robbers told them that the new hosts were only too happy to pay so dearly to have great gods among them as guests. Anyway, they would be amply rewarded by the next tribe, who would pay the same price or more. Neither *Shipwrecks* nor Oviedo manages to convincingly describe or adequately explain this strange interaction between the different Indian groups. Both accounts reveal the Spaniards' failure to understand what was going on, and modern readers are also left with a puzzling lack of clarity about why this happened.

THE SURVIVORS WERE now not far from Pánuco, perhaps as little as a hundred and fifty miles away, where Spanish raids had forced thousands of Indians to escape north. Other coastal nations had also fled their traditional lands and sought refuge from Spanish slavers inland. The survivors had walked into the midst of social chaos, made worse by western diseases that were ravaging the people.

Disease was the worst of the many tragedies caused by the European conquest of America. The statistics are uncertain and have been heatedly debated. The first scholars who estimated the population of the New World at the time of Columbus's arrival in 1492 suggested that there may have been about 1 million people in the area today covered by the United States and Canada. But then, in the 1960s, the lone voice of Henry Dobyns protested that those numbers were much too low.

Dobyns understood that previous research had ignored one all-important factor: Native Americans had no immunity to many of the most virulent and deadly old world diseases, and those diseases traveled much farther, wider, and faster than the conquistadors and colonists. The earlier estimates were always based on reports made by the first Europeans to come into contact with the various nations and tribes, but those witnesses arrived long after those tribes had already been greatly reduced by disease and the Indians' "numbers" had already "become thinned."

Of all the mighty adversaries faced by the indigenous people of the Americas during those early years of Christian conquest, by far the most destructive, by far the cruelest and most merciless conquistadors were not men at all, but the new pathogens they brought with them: influenza, cholera, typhoid, measles, and smallpox. "It may be no exaggeration that European-borne disease killed more American Indians in those first few decades [after Columbus's first voyage] than were born in the next four hundred years."

Smallpox was the most deadly of those diseases, and if it did not kill, it left deep scars. It is closely linked to damaged sight, and in Europe as recently as the eighteenth century one-third of all blindness was blamed on it.

The survivors now came across a tribe of Indians who were severely afflicted by blindness. Many who were not actually blind had cataracts, struggling to survive as partially sighted hunter-gatherers. These Indians were refugees from the south who had escaped the Spanish slavers but had not escaped disease.

"There," Oviedo wrote, with evident ironic incredulity, the survivors "cured all those who were blind or had cataracts and cured many other illnesses. Or, at least, if the Christians did not heal all of them, the Indians thought that they were able to do so." *Shipwrecks*, uncharacteristically, makes no similar claim.

Both accounts give the impression that the survivors were genuinely shocked and dismayed by the plight of the blind Indians, but we can assume that, in reality, there was little they could do to help. They moved on again, as before, making brief visits to various family groups or bands belonging to the same Indian tribe or nation. They were revered and well fed, but exhausted by the endless night dances and the strain of performing their healing rituals on so many unfortunate people.

NOW, THE GREAT 3,000-FOOT ESCARPMENT of the eastern Sierra Madre rose up ahead of them, an unassailable natural wall running almost north-south that dominates the land west of the Rio Grande. This vast natural barrier should have decided the survivors to turn south. They would have been utterly foolhardy to climb up into the mountains and so it was only by going south that they could hope to reach the coast and Pánuco. South was the way home.

Overnight, crowds of Indians had traveled from the south to pay homage to the survivors. A crowd gathered on the banks of the nearby river and beseeched the four to go back with them toward the coast. And their hosts urged them to do so as well. The way to Pánuco was opening up for them.

Yet over the next few days they were to take an extraordinary decision. It is not clear at which point the survivors turned away from Pánuco and the hope of reaching Spanish civilization. But they did. They reached a river, and instead of following it downstream, they went up it, away from the sea. It is impossible to identify which river they were on or what their route may have been, so at first they may have followed a stream that flowed southwest. But they soon turned north, heading away from their hoped-for salvation.

"Why?" asked Mendoza and his magistrates at the Mexican Audiencia in the summer of 1536.

"Because we were terrified of the coastal peoples." Oviedo reported the reply, while *Shipwrecks* recorded that they preferred to travel inland, where the Indians treated them better.

"Why?" Charles V and his advisers in Spain asked Cabeza de Vaca in 1537.

"By traveling through those lands we intended to observe them and document their characteristics," Cabeza de Vaca now claimed, hoping to sound dutiful but in fact being unconvincing.

The gulf coast that would lead the Spaniards home could only lie to the south. They were faced by a wall of impassable mountains in the west, blocking any inland route that might run parallel to the coast. It should have been imperative not to stray very far from the sea, even if they may have the coastal tribes because of their experience among the Karankawa. In fact, they had every reason to believe the coastal Indians as far as Pánuco would prove benign, for the people from that area who now urged them southward clearly revered them as supernatural beings.

By contrast, the road north could only lead them farther from Spanish lands. For seven years, they had roamed in unknown Indian territory, lost and almost without hope of escape. With hindsight, Cabeza de Vaca could tell his emperor of the Spaniards' fearless decision to further explore those territories and Indian nations for the good of Spain and the empire. But in northeastern Mexico, in the fall of 1535, this noble thought can have been in nobody's mind, for the Spaniards can have been concerned only to find a way home.

Under the circumstances, it is impossible to believe that Cabeza de Vaca, Castillo, or Dorantes would have chosen to go any other way than south. But for some reason they now traveled relentlessly northwest and then north, far into a land that was apparently unknown even to their Indian hosts. They flirted with perdition. The Indians told them that there was nothing in that direction: no people, no prickly pears, nor anything else to eat. It would be a futile adventure that would almost certainly end in a dry death, deep in the desert.

Even so, the four now sent two Indian scouts along the northern trail in the hope they would come across another tribe or nation—any people they could find, be they friend or foe. But without waiting for this search party to return, the following day the survivors set out themselves with an entourage of women water-carriers and an army of devoted braves.

Two or three hours into their journey, they came across the two scouts, who claimed to have traveled a long way before turning back, having found no one and nothing. Again, the Indians tried to persuade their guests to change their minds, but to no avail. The shamans had decided on a different course, and the Indians, weeping, now took their leave of their willful guests.

The die was cast.

The survivors themselves shouldered heavy burdens of food and water and continued upriver.

No one has ever established a satisfactory explanation for this bizarre decision. Perhaps this should not be surprising, given that neither Oviedo nor *Shipwrecks* offers a plausible motive. In truth, this marked change in direction is inexplicable when examined from the point of view of the three Spaniards. It defies logic. It is madness without method.

But Esteban had a very obvious and understandable motive to turn away from the logical road to Pánuco. The former slave was now a revered shaman. For the first time in his life he felt the easy strength that power and glory can bring a man. His healing hands held sway over the fate of other men; his breath was the breath of life. He seemed to command Death herself. And while the Indians were cowed before his godlike presence, his former Spanish masters respected him as an essential agent of their own power and salvation. For Esteban, Pánuco, Mexico, or Spain held the prospect of chains and slavery, whereas here among the Indians he was a living deity.

We are faced with a conundrum: either the Spaniards had some good reason to prefer the northerly route which has gone unrecorded and will probably never be known; or the decision to turn away from the south was taken by Esteban, and his companions preferred to follow rather than press on to Pánuco without him or try and force him to accompany them southward.

As the four men followed this northerly path away from Pánuco, they soon saw two strange, distorted figures approaching them in the opposite direction along the trail. They must have been as puzzled by the presence of these unexpected inhabitants of this reportedly uninhabited land as the inhabitants were surprised by them. As the two groups came close, the four men saw that the strange figures were in fact two women bowed down under a heavy burden of mesquite flour. The women stopped and offered some of this food to the four strangers, and explained that they were taking their cargo downriver to Indians camped a few leagues farther on. Not far upriver, the women told the four men, they would find plenty of people and food. The scouts who had told them the land was deserted had clearly been lying.

The four continued walking all day, and as the golden sun began to glow red in the west, sinking behind the ragged, rocky ridge on the horizon, they found the Indians camped in the shelter of a copse. There was little food—a few prickly pears—but it was enough.

The survivors now knew that their hosts of the night before had lied to them. At sunrise those same hosts appeared, descended on their neighbors' camp with rapacious greed, and took everything they could find, leaving nothing behind. But there was no resistance to this strange custom, a practice that the Spaniards never understood. The attackers assured the victims that the four wise men from the east were "children of the sun who could heal the sick and had the power of life and death." They advised their neighbors "to treat the survivors with great respect and to take great care not to anger them."

Shipwrecks noted that these Indians told "other lies as well, even worse than these, which they are very good at when it suits them."

Later, adding that "all these Indian nations are much given to fiction and they are great liars, especially if they hope to gain some advantage from their fibs," Cabeza de Vaca might as well have been describing himself or one of his companions.

There seem to be as many theories about where these events took place as there are theorists. The most plausible of these place them somewhere near modern Monterrey, the rich industrial capital of northeastern Mexico. This has suited a remarkable riddle thrown up by both accounts that was seemingly solved by Alex Krieger, a keen sleuth of the route.

Krieger noted carefully that *Shipwrecks* tells us the four men now headed north, skirting the foothills of the eastern Sierra Madre, until they reached a "mountain of seven leagues," the "rocks of which were seemingly slag iron." They crossed this hill and descended to a beautiful river valley, where they came across a group of Indians who were returning to their lands in the north. These people belonged to a quite different, altogether more sophisticated tribe, and they gave the survivors many gifts. Among these were a black powder the Indians used as face paint (like kohl), and some other substance that may have been iron pyrites. There was also an abundant supply of pine nuts with remarkably thin shells, more exquisite than anything the Spaniards had tasted in Spain.

In the 1950s, Krieger was the first to identify this place as the Sierra de Gloria, near modern Monclova, where there was slag iron and a mountain of the right size. Scientists have since found a type of pine nut growing there that has an especially thin shell.

The appeal of this academic treasure hunt is unquestionable, and Krieger may well have been right. But finding the precise route is a distraction from one important detail: these Indians also gave the survivors a considerably more interesting gift, a copper bell decorated with the image of a face. Archaeologists have found many copper bells in the American southwest that date from before the Spaniards arrived. These are mostly undecorated and are very small, rarely much bigger than an M&M. It is clear that the bell given to the survivors was a much larger and more sumptuous piece. It was held in high esteem by the Indians and was of considerable interest to the Spaniards, ever hungry as they were for signs of civilization such as metal smelting.

The Indians said it came from a neighboring tribe from the north who manufactured many such bells. That provenance is borne out by the discovery of a similar bell at the ruined Hohokam pueblo near Mammoth, Arizona, which when first found was mistaken for a flattened ball cock from a lavatory cistern. It has also been suggested that the bell given to the survivors came from the Casas Grandes site in northern Chihuahua, which had been abandoned by the 1530s. Both places are well to the north of the Sierra de Gloria.

The survivors continued north, traveling among neighboring bands of hunter-gatherers. These people had "the most beautiful method of hunting you could think of," according to *Shipwrecks*. They spread out across the terrain, carrying "hunting sticks that were at least three palms in length." Then, when one of the many hares that gamboled about those fertile lowlands sprang from the scrub, the hunters advanced, surrounding their prey. They wielded their clubs with such dexterity that the hare would rush from one to another, "which was quite astonishing to see," like some primeval game of hockey. There was also a plentiful supply of pine nuts, and for days the survivors were happy, easily following well-worn trails with their bellies full, surrounded by a horde of adoring believers.

During this pleasant progress, *Shipwrecks* reported, an event took place that greatly reinforced the growing reputation of the survivors among these inland Indians. A man came to Cabeza de Vaca seeking help for an old war wound. An arrowhead was buried deep in his torso, above the level of his heart. *Shipwrecks* claimed that the ever fearless Cabeza de Vaca now produced a knife—a knife which we must assume he had managed to keep hold of during all his years of captivity among the Karankawa—and opened up the man's chest. The arrowhead had pierced the cartilage and become twisted sideways, so Cabeza de Vaca had to cut deep into his patient in order to remove it. With this operation complete, he used a bone needle to sew up the wound with two stitches, and then stemmed the bleeding with a leather compress. The patient healed quickly, and all his people came to gawp at the arrowhead and to meet "the first surgeon in Texas," as Cabeza de Vaca is remembered by the Texas Surgical Society, although this dexterous but relatively straightforward surgery is in fact described as taking place in Mexico.

Shipwrecks reported that as the news of this surgical procedure spread, it inspired reverence and love in the Indians, but Oviedo makes no mention of it.

This story of surgery seems similar to the evident embellishments encountered earlier in *Shipwrecks*, the strange tales of Evil Thing, raising men from the dead, and of the biblically burning bush. And just as Dr. Sibley offers a tantalizingly plausible explanation for that tale of the spontaneously combusting tree, he also tells a strangely similar story of basic surgery in Indian country:

> Last summer an old Spaniard came to me from Labahie, a journey of about 500 miles, to have a barbed arrow taken out of his shoulder, that one of these Indians had shot in it. I found it under his shoulder-blade, near nine inches, and had to cut a new place to get at the point of it, in order to get it out the contrary way from that in which it had entered: it was made of a piece of iron hoop, with wings like a fluke.

Without antibiotics, the task faced by Sibley at the beginning of the nineteenth century was little different from that faced by whichever survivor may have removed the arrowhead from the Indian's chest in the 1530s. It is entirely possible, therefore, that one of them did so, since the operation itself was feasible. But it is equally possible that the whole story was made up, given Oviedo's silence. Whatever the truth, which one of them may or may not have been the surgeon is, of course, a matter of pure speculation.

Now the survivors ventured farther and farther north and began climbing steadily into an arid, rugged, mountainous desert, without game, without people, while the Indian followers who remained with them gradually began to fall ill.

There is the usual lack of clarity about their precise route at this point. Yet it seems reasonable to suppose that that they were crossing the Sierra del Carmen, the dry mountains that loom above Big Bend National Park, Texas, from the Mexican side of the Rio Grande. Then they descended into the river valley and waded across the modern international border again, reaching the swath of barren plains below the Chisos Mountains, where they found a great crowd of people who had gathered to meet them.

The lunar landscape of the place made a beautifully hostile, uncanny, alien backdrop for their reception. The volcanic desert is sliced by the canyons of the Rio Grande as it finds its way between the tectonic upheavals of Big Bend. The horizons to the east, west, south, and north are dominated by the shimmering grays of the Sierra del Carmen and the browns of the fertile Chisos mountain basin. Emory Peak, the highest point in the modern National Park, dominates the northern skyline. The plain between is scarred by waterless ravines and the crumbling purple rock now perched on piles of compounded ash that were once spit out by now long-deceased volcanoes.

The people of the pine nuts who had brought the survivors out of the mountains and into the desert plain now turned for home, laden with gifts from the people of the Rio Grande. Unable to carry all that they were given, they abandoned much of it at the side of the trail. They also left behind them a deadly Old World pathogen.

THE SURVIVORS NOW turned to their new hosts and explained their intention to head west, toward the setting sun. The Indians told them that there were many miles of deserted country in that direction, beyond which there were hostile tribes. Two Indian women were immediately sent as emissaries, and Cabeza de Vaca and Dorantes both give the impression that they believed these women had orders to make contact with those hostile tribes in the west.

The survivors now insisted that their Indian hosts accompany them on the next stage in their journey, but their hosts refused. Oviedo now reported that Andrés Dorantes told one of the Indians to inform the others that their punishment for this truculence would be death "and this struck such fear into them that the following day they returned sick from their morning hunt." Within two days, more than 300 were ill—afflicted, we must assume, by whatever disease the people of the pine nuts had brought with them across the mountains.

The Indians were terrified, convinced that the angry Christians were causing the sickness. For fifteen days they cowered in silence, not daring even to eat or drink without permission.

Shipwrecks, as has already been explained, tells the story with a small but all-important difference. Angered by the Indians' intransigence, Cabeza de Vaca, *Shipwrecks* claims, left the encampment and slept out in the desert, under the starry sky, well away from their hosts. The Indians went out to see him and told him how frightened they were. They begged the survivors not to be angry and promised to take them wherever they wished to go.

Was it, in reality, Esteban's sleight of hand that turned an unhappy epidemic into a powerfully persuasive ruse?

There is something thoroughly suspicious about the way the two accounts describe this crucial juncture in the four men's story of survival. Both Oviedo and *Shipwrecks* state that the survivors were determined to travel west. Yet in both accounts they in fact went north, following the almost deserted valleys of the Rio Grande as far as El Paso, where a great Indian highway approached the river across the desert by way of the watering holes of Hueco Tanks. Only then did they turn west, where the well-worn trail made the crossing, *el paso*, below the modern city that gets its name from this ancient ford.

Once again, the accounts offer tantalizingly inconclusive details which suggest Esteban caused this change of direction.

When the two women emissaries returned, it was immediately obvious that they had gone north—not west, as Cabeza de Vaca and Dorantes seem to have believed. They brought news that the semisedentary people who lived farther up the Rio Grande valley had migrated north to hunt bison on the plains, abandoning the lands through which the survivors were hoping to travel. That news later proved inaccurate, but it was enough, according to both accounts, to convince Cabeza de Vaca and Dorantes that they had to try to find a route west through the mountains. However, it is also obvious from both accounts that Esteban and Castillo did not agree and that the rift was temporarily irreconcilable; unbelievably, given the circumstances, the survivors now split up.

As so often happens, the accounts do not coincide; nor is there much clarity about what really happened. But Cabeza de Vaca and Dorantes either remained where they were or began to travel westward. Meanwhile, Esteban and Castillo set out northward in search of the bison hunters and took the two women emissaries with them as guides.

For three days the women led Esteban and Castillo through the beautiful, empty landscape. It was a long march that took them far away from their Spanish companions. Finally, they reached a settlement where the father of one of these female guides was chief. This settlement completely changed their view of the world. The people lived in permanent dwellings, the first that Esteban and Castillo had seen that "looked like real houses." They planted crops, albeit in rudimentary fashion. This was the first farming community that the survivors had seen in the eight years since they left Cuba. It was an oasis of green in the desert of dusty hues, inhabited by "the best physical specimens they had seen, skilled and sophisticated people who seemed to understand the survivors better," than the other Indians did. It must have seemed like some Elysian field even though it was merely a few shacks on the banks of "a river that ran between the mountains."

This garden paradise was perhaps near the Santa Elena Canyon, where the Rio Grande flows with sinister gentility between looming walls of rock; or in one of the valleys upstream from Lajitas, a rather peculiar, stage-set town that nowadays is the home of a golf course which advertises a ninth hole with the tee in Texas and the green in Mexico.

Esteban and Castillo were overjoyed to be amid fields again and to see soil broken by human endeavor. There were beans and squash and, most important of all, these people had corn. They carried water in gourds and offered their guests many bison hides. They told them of larger settlements a short journey up the river, where the Spaniards would also be well received.

Whatever had caused the four survivors to go their separate ways, Castillo knew that it was the duty of a nobleman and a Spaniard to take the good news of this semi-civilization to his companions. He set off immediately with a handful of Indians, in search of Dorantes and Cabeza de Vaca.

Shipwrecks presented this brief rift between the survivors as a planned attempt to persuade the bison hunters to guide them along their chosen road. In that version of the story, Castillo and Esteban were merely messengers, sent to bring the Indians who had gone north back to the survivors' base camp among the Indians gathered on the Chisos plain. But the truth was that as soon Dorantes and Cabeza de Vaca learned about the farming village, they abandoned any alternative plans, evacuated their base camp, and went with Castillo to rejoin Esteban.

Meanwhile, Esteban remained among the bison hunters, always on the frontier between one culture and another, always far from home, but always, it seems, at ease. He must have contemplated staying with these friendly people. Theirs was an attractive world, a life lived between the relative comfort of that bucolic paradise and the dangers and rewards of the bison hunt out on the exposed plains. Compared with the hardships Esteban had suffered among the coastal Karankawa and the deadly poverty of the convivial Avavares and their friends, this place must have seemed like Eden, despite the fact that the Indians claimed "it had not rained for two years and the drought had lasted long enough for all of them to lose their corn crop."

"These people were the best looking, the most vivacious, and most advanced" that the survivors had so far encountered. "We had a better rapport with them and their answers to our questions made more sense. The men all walked about naked, while the women and some of the old men wore deerskins." Esteban must have wondered what life might be like among these buffalo hunters, deep in the heart of America. Had he found a happy island in that desert, where a slave might be forever free, a place where he might satisfy his natural desires, a humble but safe world, where there were no Christians to thirst for gold? Might not this land of Indian braves make a happy home in which to rest at night below the star-spangled firmament and enjoy a true sense of liberty?

But Esteban, now alone among these people, even as he asked himself these questions, must have known that this was no place to abandon his Spanish companions, for all that he might revel in brief contemplation of that wistful dream of a forgotten future. Life as a black slave among the Spanish conquistadors was an evil business, he knew. But he also knew that this lonely place, beyond which neither African nor European had yet ventured, was not Eden but instead a home to poor Indians who saw their god in the clouds and heard him on the wind. A life of noble savagery is perhaps better left to poetry than lived by man himself.

Surely Esteban recognized that in reality he was merely resting at one of a few sedentary villages on the edge of a vast desert plain that would take "thirty or forty days to cross." Esteban was a metropolitan character. He had lived in Azemmour, in its Portuguese heyday. He had lived in Seville, the third largest city in Europe and certainly the most cosmopolitan. He knew the busy towns of Cadiz and Jerez. He had been to the bustling port of Santo Domingo.

He was now among people who boiled water by heating stones in a fire and throwing them into a cooking pot made from a dried pumpkin—ingenious, but primitive. Their simple houses only reminded him of the great cities he had known. Their cooking whetted his appetite for a future feast.

With his easy charm, Esteban carefully cast his spell of shamanism over these isolated farmers, curing the sick, taking part in their dances and ceremonies, urging them to warmly welcome the white gods who would soon be among them. Now that he had gained their trust, he gathered all the inhabitants of the village together and led them back along the trail that went down onto the plain, there to meet the three Spanish witch doctors.

What natural and understandable hubris must have filled the free heart of this erstwhile slave as he triumphantly led his army of buffalo hunters to come face to face with the three European aristocrats and their rabble of diseased, uncooperative followers. When the two groups met a few miles along the trail, Esteban's people brought farmed food and water to feed the three hungry Spaniards and their worn-out entourage. This was a victory for Esteban's conviction that the right route was to the north. It confirmed him yet again as the leader of the Spaniards.

The four men now dismissed the unhealthy Indians who had traveled north with the three Spaniards, rewarding them with the surplus gifts they had received from Esteban's bison hunters. Those now disillusioned, diseased, and dying people made their way back to their camp, their world contaminated by the passing magicians.

That night, the four survivors celebrated with their new hosts. They ate and drank; they partied. The Indians ranged themselves around a great fire and made music, clapping and singing. They danced to the rhythm of haunting voices, chanting *Ayia canima ayia canima*, and they wailed to the beat of a drum made with "skins attached to a vessel in the form of a tambourine." The naked dancers on each side of the fire took turns standing up, in pairs or sometimes four or eight at a time. They then began to "whirl around like jesters, raising their hands towards the sun and singing." They performed with such striking "unity and harmony" that a later Spanish expedition reported, "though there be three hundred savages in a dance, it seems as if it were being sung and danced by one man only, due to the fine harmony and measure of their performance." To the Spaniards this dance "resembled that of the *negros*." Esteban's opinion is not recorded.

THERE ARE PUZZLING omissions in *Shipwrecks* and Oviedo, striking characteristics of the American South that go unmentioned. It is, for example, inexplicable that there is no mention of alligators in the swamps and marshes of Florida. It is easy to understand, however, why the survivors were silent about peyote, the potent natural drug that was often taken by Indians as part of their religious rituals. Peyote is a strong mix of stimulants and sedatives; it causes hallucinations and suppresses hunger. Shamans used it in performing magic and in curing disease. There can be no doubt that the four survivors took it frequently as they made their way north curing the sick, for the Rio Grande is in the heart of peyote country. But it was a characteristic of Indian culture that the Inquisition and Zumárraga would have viewed with considerable suspicion. It would have been unwise to document their experience of peyote in case such evidence was turned against them.

As the four men reveled in their temporary felicity, Cabeza de Vaca and Dorantes knew well enough that they had Castillo and Esteban to thank for their lives. It soon became clear that their proposed route to the west would have taken them along arduous mountain trails far into the Chihuahua highlands. Nature's great defenses would have blocked their path. They surely would never have seen their homeland again had they gone that way.

Instead, the following morning they continued up the Rio Grande, traveling through the fields and gardens of the cultivated floodplain, going from settlement to settlement, their bellies full. They were now among Jumanos Indians, and the people of each village waited for them in their houses, their heads turned to the walls, their hair combed down over their eyes. They offered their worldly goods to the four survivors, piling their possessions in the middle of their living quarters as an offering and a ritual of homage. The Spaniards were most struck by the quantity and quality of the buffalo hides they were given. At least, that is what they told the Mexican Audiencia in 1536. But perhaps their anxiety to report something of value coupled with the poverty of their pathetic existence during the preceding years, led them to remember riches where the magistrates would have merely seen just a few paltry animal skins. Wealth is relative and beauty is in the eye of the beholder.

The reality of these gifts was as valueless as the account of them given later in Mexico, for the survivors were forced to leave the hides behind. They themselves could carry no more than the bare necessities, and these Indians had no interest in accompanying them on the next stage of their journey. Their annual migration to hunt buffalo on the plains was urgent business. Food was becoming scarce, the Indians explained; they were now in the second year of a drought, and without rain they could not sow their corn. They had lost their crop the year before, the land was sick, and they feared famine might strike soon.

The survivors wanted to know where the corn they had been eating had come from.

It came from far away, they were told, toward the setting sun and to the north. The place was a journey of thirty or forty days among desperately poor peoples who had no beans and little else to eat. The way to reach that land of corn was to follow the river north for nine or ten days, to the place where the great trail from the plains crossed the stream. They should then follow that trail to the west until they found the corn lands, which stretched south, north, and west as far as the sea. These were the sparse but functional directions that the Indians gave them.

To the survivors, corn signaled cultivation and agriculture, signs of civilization. The four men now had a new goal, one that they held in common. They were determined to reach the land of corn, and there was little, it seems, that could prevent the progress of these hardened survivors and they took leave of their hosts and set out on the northern road.

Half a century later, two large Spanish armies marched through these permanent settlements of the Rio Grande within a year or two of each other. The Indians warmly welcomed the men of both expeditions and told them that many years before, four other strangers had passed that way. The chroniclers of both these later forays described this area of the Rio Grande valley around La Junta de los Rios as fertile farmland, where the Spaniards ate well. There were beans and squash. There was plenty of maize and mescal, and they were offered gifts of bows and arrows and buffalo skins.

Dorantes reported that when the four survivors marched upriver for fifteen days without respite, they traveled among people who had almost no food to give them. Most of these Indians had gone to hunt buffalo. But the four men were long since inured to such privations, and they survived on a handful of deer fat and "something those Indians called Masarrones, which they gather from the trees." These were wild mesquite pods, which one of the later chroniclers described as being like the screws on an harquebus, small and yellow, with pits like a prickly pear. The Indians ground up these meager beans, but, Dorantes explained, this produced nothing but splinters, which made for very bad food, even for animals.

As they traveled up the Rio Grande, the friendly Jumanos Indians must have told the survivors that there were larger and richer settlements to the north, just as the later expeditions learned that "thirteen days travel upstream," they would find people who "wore clothes and grew and harvested plenty of corn, calabashes, and beans. They also grew cotton which they spun, wove, and made into blankets and shirts with which both the women and men clad themselves."

But according to *Shipwrecks* and Oviedo, somewhere near El Paso, rather than go in search of these settlements of the upper Rio Grande, the four companions preferred to turn west. Both accounts suddenly give little or no information. We are simply told that the survivors passed among Indians who hunted hare and ate "powdered grasses" or "straw." They walked for seventeen or twenty days. Occasionally they rested. They were accompanied by Indians. We are told no more.

Instead, we are rushed headlong through the story until the survivors reach a "land of maize," the fertile valleys of the Opata Indians, with their corn-filled fields, cotton clothing, and permanent houses built of mud and woven cane. There the survivors feasted on venison and toasted corn tortillas. They were regaled with fine cotton blankets, beads, and corals from the South Sea. They were given turquoises from the north and five fine emeralds from the high mountains, each stone worked into a sacred, ceremonial arrowhead, bartered by the Opata in exchange for plumes and parrot feathers.

"Among these people," Cabeza de Vaca remembered, were the most "modestly dressed women," they had seen, who "wore cotton blouses, with long leather skirts, and shoes."

Here the wanderers turned south toward the sea. They descended through the Sonora valley, still today a garden paradise in the desiccated world of northwestern Mexico. The life-giving river flows year round through the sleepy modern farming communities, bringing a cool current of clear water by which families picnic on the banks, and lovers linger into the night. Shade trees frame the hot, dusty plazas, sheltering the corn and taco sellers from the mid-morning sunshine.

Esteban, Castillo, Dorantes, and Cabeza de Vaca passed through these oasis towns and broke corn bread and tortillas with the predecessors of today's farmers with their starched Stetsons and battered pick-ups. They passed through Arizpe, which then had 600 houses and seemed to them like a great metropolis.

For the first time in the account they gave of their survival, the four wanderers described a world that was recognizably civilized to European eyes. They were now among a people who had agriculture, temples, and towns. These towns were well defended by fortifications constructed in the hills above. Most important, these Indians were part of a trading network that stretched far to the north and south. These were the first clear signs of wealth and good living that the survivors had seen since they set sail from Cuba in 1528, eight years before.

But it was also a place of brutal religious ritual. Just north of modern Baviácora, archaeologists have found the ruins of a great public architectural project. Two slender platforms, 150 feet long, form the sides of an enclosed plaza, seventy feet in width. These may be the remains of the great stone temple of Chimcamatle, described by Las Casas. It was topped with a statue or effigy, hung with dried animal hearts and coated with the blood of sacrifice.

The lords of the land climbed the steps of this temple and called out to their people like town criers or Muslim muezzins, urging them to hail the strangers. The Indians carried a great idol ahead of the four survivors, leading them to the temple. There, to the melodic sounds of flutes and pipes, they sacrificed hundreds of animals. Deer, hare, wolves, and birds were "split open and their hearts were ripped out, and they bathed the idol in the blood that gushed forth. The people prostrated themselves in obedience to their pagan gods."

The Sonora River turns abruptly west below modern Baviácora to cascade through a steep canyon, before flowing gently into the sprawling city of Ures. Somewhere near here, the four wandering witch doctors spent Christmas 1535 at a group of three small settlements of twenty houses each. They were finally out of the mountains and onto the seaboard, and they were welcomed by the people of these villages, who brought them more than 600 desiccated deer hearts as a sign of reverence. They named the place the "Town of Hearts," Corazones.

Castillo was the first to notice the Spanish scabbard buckle and horseshoe nail worn as a necklace by one of their Indian hosts. Taking it in his hands, he asked about it as best he could.

"What is this?"

"It comes from the skies," the Indian indicated.

"From the heavens?" Castillo asked. "Who brought it from heaven?"

"Men like you, with beards, who came down out of the sky to this river."

Cabeza de Vaca, Dorantes, and Castillo rejoiced and gave thanks to the Lord for the first news of fellow Europeans that they had had in eight years of isolation. But their joy was short-lived and bitter.

"What happened to these men?" the survivors asked.

"The men from the sky went to the sea, where they had put their lances into the water, and then they themselves also went into the water, then under water. But later they reappeared on the horizon at sundown."

The four men knew then that the Spaniards had sailed away. There, on the Yaqui River, one old history tells us, the survivors sat down and wept over their long journey and their exile among the Indian nations. But as they shed those tears, an "Indian of most civilized bearing" asked:

"O gods, why are you sad and unhappy?"

They explained and the Indian spoke again.

"Nearby," he said, "there are many other men like you, who ride around on swift animals," wreaking havoc wherever they go.

Such certain news of possible salvation so close at hand spurred the four survivors to press onward at greater speed. But we might pause to wonder what Esteban felt.

One Spanish conquistador, a man who knew Esteban, reported that he always wore his shamanistic finery: a splendid feather headdress, and jangling bells strapped to his arms. He carried his shaman's scepter, a richly decorated gourd, the native rattle from near Monterrey in Mexico which was a symbol of his power and which sounded like the sinister drone and cackle of an angry rattlesnake. He had a collection of fine ceramic plates decorated in many colors, no doubt personal souvenirs. Another historian described how Esteban developed friendships with the people of northwestern Mexico, so much so that they pressed upon him gifts of turquoises and other valuables. They even offered him their womenfolk, whom, that lewd commentator observed, he very much enjoyed.

What hope, then, did Esteban have for his future when his companions were so overjoyed at finding evidence of their fellow countrymen close at hand? The Spaniards were headed for home, for the Spanish world. But where was his home now? In his childhood world beyond the Sahara, or in far-off Azemmour? For eight years he had lived as an equal of highborn, high-handed Spanish aristocrats. Could he really return to servility? Might slavery in the material wealth of the European world prove preferable to the spiritual splendor of life among Indians who treated him like a god? Why should he choose the Old World over the New and prefer European ways to American customs? Whatever his reasons, he chose to return to the Spanish world.

THE FOUR SURVIVORS pressed on along the trail, and as they traveled they heard ever more disturbing tidings of the Spaniards they were searching for. The Indians had abandoned their fertile, well-watered land of rivers and had fled to the highlands or gone into hiding deep within the bush. They were too afraid to work their lands and live in their villages for fear the Spaniards might return and capture many of them as slaves. They were hungry, living on the bark and roots of scrubland trees. They were grotesquely thin and weak from disease.

The four companions shared the Indians' suffering as the Indians told the same story again and again. Strange foreigners had invaded their lands and destroyed and burned their settlements. They carried "sticks of fire" and rode on enormous "deer." They had taken away men, women, and children, tied together like rabbits on a stick. The four horrified survivors were distraught. They vowed that they would catch up with the Christians and order them to cease the killing and enslaving. They sent out a rallying call, which many Indians answered. They began to build an army of allies as they hurried along the back trails of the bush and climbed up a steep and rugged path to reach a settlement perched high on a mountain ridge. The people of this overcrowded town welcomed the four survivors and gave them 2,000 loads of corn, which they quickly distributed among the hungry men and women in their army. High up and isolated and well stocked with food, the place seemed a safe refuge from the Spanish terror. But it was no place to linger long, merely a place to regroup while the four survivors quickly planned their next move.

They had to act fast if they were to get home, and so, as they had often done before while traveling across America, they sent out messengers with orders to spread word that four great shamans were coming. They named a trysting day and summoned the scattered Indians to a settlement three days' journey from their mountain fastness. The next morning, the four men set out with a vast company of companions and headed back down out of the mountains. As this makeshift fighting force traveled, fleet of foot, through the bush country of the high seaboard, they came across campsites where Christian slavers had slept.

At midday, their messengers appeared, returning along the trail with unwelcome but not surprising news. There would be few new recruits at the mustering place because the inhabitants of that country had fled into the bush. Soon afterward, Indian spies came to them with alarming news of slaves they had seen chained. Panic spread easily through the ill-disciplined array. Many men and women fled the makeshift cohorts and rushed back along the trail to warn their relatives and friends of impending doom. But the four survivors managed to persuade many others to stay, reassuring their people that that they need have no fear.

History does not record whether Diego de Alcaraz, Lázaro de Cebreros, and their men had been aware of the Indian spies hidden in the undergrowth and trees of the *monte*, the impenetrable bush that hemmed them in. But, with the experience and cunning of hunters at home in their ancestral lands, these Indians no doubt went unnoticed as they tracked the slavers' slow progress south.

The following day, the spies led the four strangers and some of their newly gathered forces to the place where they had seen Alcaraz and his men. The four survivors knew at once that there were many cavalrymen amongst the slavers. They found the places where the horses had been staked and tethered and saw the other detritus of a fast-pitched and fast-struck camp.

Cabeza de Vaca asked his readers to imagine the gladness that swelled in their hearts when they realized just how close these other Spaniards were. With almost biblical rapture, he wrote of how the four gave thanks to God, who in his mercy had granted them salvation from the wilderness where they had suffered such hardships and such fearsome dangers. And so, Cabeza de Vaca said, he suggested that one of his companions should go on ahead and catch up with the slaving party. But his companions thought it was a bad idea. They complained about how tired they were and how much effort was involved. Consequently, even though Dorantes and Castillo were much younger and tougher, it was Cabeza de Vaca who, the following morning, took "the black man and eleven Indians" and set off in search of salvation. Cabeza de Vaca is ever the hero of his own story, but Esteban is never far away.

Esteban and Cabeza de Vaca moved fast, racing along the trails, easily following obvious tracks made by Alcaraz and his men.

And here we reach the point where this book began.

Part Five

THE SEVEN CITIES

OF GOLD

1536-1539

The Gulf of Mexico (c. 1544). This slightly later map shows the River and Bay of the Holy Spirit and the "Bahía Honda" (Tampa Bay, the Deep Bay) on the gulf coast of modern Florida. *(Courtsey of Archivo General de Indias; MP México 1)*

18

FRIAR MARCOS DE NIZA

1536–1537

IN PART ONE, we traced the survivors' journey from north of Culiacán, in the far northwest of Mexico. We saw them meet the Spanish slavers and Melchior Díaz, before traveling on to Guadalajara and the viceregal capital, Mexico City. There, they were royally treated by the viceroy, Antonio de Mendoza; and by the great conquistador Hernán Cortés. They also met the stern Archbishop, Juan de Zumárraga, who had his own plans. These mighty Mexican oligarchs tried to bribe and cajole the four survivors to ensure that the official report they made in the Audiencia served their own powerful interests.

Oviedo and *Shipwrecks* give the modern historian a good but imperfect idea of what went into that first official report made in the summer of 1536. But we also have a strong sense of the exaggerated and fantastical rumors that quickly developed as the four survivors embellished their own stories in private, in the taverns and at the dinner tables of Mexico City, telling their tale in the way that adventurers are prone to do. And it was not simply that the survivors told the tale again and again, but, then, again and again it was retold by others, changing, developing, and all the time becoming more fanciful. With each new telling another layer of fantasy further veiled the truth beneath.

During their long march across America, the survivors had learned about the large Indian towns on the upper Rio Grande from the Jumanos Indians of the area around Big Bend and the Junta de los Rios. These belonged to an archipelago of important settlements that made up the sophisticated Pueblo Indian culture, which thrived as far away as Zuni on the borders of New Mexico and Arizona and among the Hopi and beyond. The Pueblo Indians lived in well-defended urban centers with large honeycomb constructions of multiple stories, the low-rise apartment blocks of a bygone world. Their society embraced politics and civic administration.

The Pueblo Indians were far more sophisticated than their neighbors, hunter-gatherers like the Karankawa or the seminomadic Jumanos. So, when the Jumanos told the four survivors about large towns to the north, they described places that they thought of as great cities of almost unimaginable size. The four survivors themselves had been impressed by the relative sophistication of the tiny permanent settlements around Big Bend, after their brutally simple life on the Texas coast. The mysterious pueblos to the north thus became a symbol of civilization for Esteban, Castillo, Cabeza de Vaca, and Dorantes while they were living as Indians.

In reality, the pueblos were little bigger than villages, no more than simple sedentary communities compared with the advanced civilizations of Aztec Mexico, Inca Peru, or Spain. But in the minds of the four survivors they had become exaggerated, because they had offered some semblance of hope in a time of desperation.

The Spaniards of Mexico City were eager to believe those exaggerations. There was a simplistic logic to the belief that if the great Inca civilization had been discovered in Peru and Cortés had found the Aztec Empire in Mexico, then it stood to reason that a similarly golden prize lay somewhere in North America. It is worth remembering that when Cortés first set foot in the Aztec capital Mexico-Tenochtitlán, he told the Emperor Montezuma that the Spaniards suffered from a disease of the heart that might be cured only with gold.

That belief in a civilization in the north was encouraged by the medieval Spanish legend of the "Seven Portuguese Bishops," who had fled from the advancing Moors during the bleakest period in the history of Christian Spain. During that Dark Age, the bishops took to their boats and sailed out into the Atlantic Ocean, until they eventually reached a great island, where they settled and founded seven Christian cities. This old legend was given a new lease of life in 1448, when sailors from a Portuguese ship claimed that a ferocious storm had driven them onto the shores of that legendary island. They said that they had been carried on the shoulders of the population to a church, where a Mass was said. Then, when the storm relented, they set sail and went home to Portugal, where they were severely scolded for failing to record the precise position of that miraculous Christian island.

North America was quickly identified in the collective imagination of Spanish Mexico as the island where the "Portuguese bishops" had established their colonies. That speculation was fueled by a story told to Nuño de Guzmán by a Tejos Indian, who reported that great civilizations lay to the north of Mexico. When he was a child, that Indian said, his merchant father had often traveled through the Pueblo country, trading his beautiful feathers for gold and silver, which were common metals in those parts. Once or twice he himself had gone too and he remembered visiting towns so large that they seemed comparable to those in Mexico. There were seven such cities, and each was home to a whole street of goldsmiths. The best way to reach those seven cities, the Indian went on, was by way of a grass-covered desert that lay between the Gulf of Mexico and the Pacific Ocean.

Long before the four survivors arrived in Mexico in 1536, the Spaniards had developed a varied but always hopeful geography of North America. In 1527, as Narváez prepared to set out for Florida, a Spaniard well experienced in the withering wonders of the New World had lampooned the ludicrous nature of Narváez's enterprise.

How, he had asked his sovereign, was one expedition to even explore, let alone conquer, the vast landmass of over 2,000 leagues that lay between Florida and the Pacific coast? But this critic, who at first seems quite sensible, later strongly advised that it would be better for Narváez to go in search of a city called Coluntapan, where the people wore silver armor and carried metal weapons. He too had fallen victim to the Spanish disease of hopeful delusion.

Spanish Mexico vibrated to the mesmerizing rhythm of such stories, which filled raconteurs and audiences alike with hopes and desires for future adventures. In 1536 and 1537, Esteban, Castillo, Cabeza de Vaca, and Dorantes sowed the seeds of a more focused covetousness on this very fertile ground and greed suddenly gripped Mexico, taking control of the collective colonial mind. The survivors told a story that easily mingled exaggerated fact with outright fiction. And as dreams turned into belief and belief took on the certainty of faith, the road to the Seven Cities of Gold came to be revered by many a conquistador as a certain path to wealth.

By contrast with the many hopeful dreamers, Mendoza, Cortés, Zumárraga, and the four survivors were pragmatists who must have guessed that rumors of the Seven Cities of Gold were simply a myth. But they also understood that such dreams and hopes were a spur to action. The Seven Cities were a metaphor for an intangible prize, a symbolic goal; but these practical men all recognized that a very real, very tangible prize was at stake in North America. The four survivors had described a "land of cows" and had explained in private that the Indians talked of a great grassy sea, endless fields where enormous herds of wild cattle roamed free. They had heard about the Great Plains. That was the news which excited Mendoza, for he had arrived in Mexico with clear instructions from the Crown to develop stock raising in the colony.

The richest and most powerful institution in Spain, after the Church and the Crown, was the Mesta, which controlled sheep farming and the wool trade. In 1537, as Mendoza and Zumárraga wrangled over the right to organize an expedition to the Seven Cities, the viceroy established the cattle breeders' association of Mexico and called it the Mesta too. The four survivors' reports of vast herds of cattle ranging on boundless grazing land were much less romantic than the thrilling news about cities filled with streets of goldsmiths that excited impoverished and quixotic conquistadors, but any rich Spanish aristocrat knew that stock was a surer path to wealth and power.

BY THE LATE summer of 1536, the four survivors had given their formal testimony and were able to relax and begin to plan their futures. *Shipwrecks* reports that as soon as they had finished with the official account, Cabeza de Vaca decided to seek a personal audience with Charles V. He set out in October for the port of Veracruz, but a storm destroyed his ship before he could embark, and he went back to Mexico City.

Over the winter of 1536 to 1537, Mendoza time and again discussed a possible expedition to Cíbola with the survivors. Meanwhile, Cortés tried to persuade the viceroy's right-hand man, Francisco Vázquez de Coronado, to lead an expedition into the lands to the north. But Mendoza was firm in his response. "You know only too well," he told Coronado, "for you know the kind of man I am, that I will never accept any such proposition from Cortés or anyone else other than His Majesty, nor would it be right to do so." Mendoza concluded by informing Coronado that as soon as he was given permission to do so by Charles V, he would personally underwrite the expedition himself. "Never again speak to me of this matter!" Mendoza commanded. The viceroy, Cortés, and Zumárraga all hoped to persuade the four survivors to guide an expedition into those rich northern lands.

But the Mexican oligarchs were to be disappointed. In February of the following year, 1537, the viceroy had to write to the Spanish Crown with the news that both Dorantes and Cabeza de Vaca would soon come to Spain to give an account of their adventures in person. There was little chance now that an expedition to Cíbola would be led by one of the senior captains who had survived the Narváez expedition.

Mendoza must have then considered Alonso del Castillo but for some reason seems to have discounted him as a possible candidate. And although no explanation for this is offered by the contemporary documents, we can hazard a guess.

We know Castillo was very young at the time he went to Florida, for, in 1547, when giving legal testimony in Mexico City, he gave his age as "35 years or more." During that period in history, few individuals knew precisely how old they were, and dates are often imprecise in this kind of document, so he may have been as old as forty or even a year or two older. Even so, Castillo would therefore have been between fifteen and twenty-one when he set out with Narváez. By the time he reached Mexico City in the summer of 1536, he had spent well over half his adult life as an Indian. He must inevitably have seemed somewhat weird if not actually mad. Mendoza perhaps felt that Castillo was mentally unstable and therefore an unreliable military leader. Instead, Mendoza arranged for Castillo to marry a wealthy widow, and he settled down for a while. Within a few years he had established a family home in the town of Puebla de los Angeles, where his wife was a landowner.

That left only one man as the possible leader for the proposed expedition. Mendoza wrote to Charles V and explained that he had "bought a black slave from Dorantes," a man called Esteban who would make a good guide for the expedition because "he had been to those lands" and "was a civilized, intelligent person."

There is some doubt as to whether Mendoza actually bought Esteban or whether he remained Dorantes's property. A Spanish chronicler called Baltasar Obregón reported that Andrés Dorantes was "noticeably upset" at the prospect of having to sell Esteban to the viceroy. "He would not sell Esteban," even "for 500 pesos," which Mendoza sent him "on a silver platter." The average price for a slave during that period was between 100 and 150 pesos, so the offer was not excessively generous if we take into account Esteban's priceless knowledge of the north. Not surprisingly, Dorantes seems to have refused. But to defy a man as powerful as Mendoza was dangerous. If pushed, the viceroy could invent some legal ruse and "steal" Esteban without much trouble. So Dorantes had to offer some kind of deal. Obregón reported that he then agreed to hand over Esteban to the expedition without charge for "the good of His Majesty" and to help save the souls of the many "natives of those provinces."

The touching description of Dorantes as "noticeably upset" at the prospect of being parted from Esteban has led some to draw a sentimental picture of the relationship between master and slave. The good Dorantes is thus pictured as effectively giving Esteban his freedom so that he could join Mendoza's enterprise. But the reality was surely more prosaic. As long as Esteban remained his property, Dorantes could stake his own claim to any riches his slave might discover: precisely the point made, in the context of writing history, by Richard Wright. There may have been a strong emotional bond between the two men after their years together in the wilderness, but Dorantes's reported refusal to sell Esteban to Mendoza was at least in part a commercial decision.

However, while Cabeza de Vaca reached Spain in the summer of 1537, Dorantes spent ten months lost at sea on a ship captained by the criminally unscrupulous Sancho de Piniga, as explained in Chapter 10. When he finally returned to Mexico, late in the year, Mendoza immediately offered him command of the expedition to the Seven Cities. But Zumárraga then forced Mendoza's hand, insisting that there should be no Spanish military expedition. Instead, Marcos de Niza was put in charge and Esteban was appointed as his guide and given the command of an army of Indian companions.

Mendoza can have only guessed at what kind of man Esteban might turn out to be, for he could have had no idea how such a unique cultural hybrid might react to being given official responsibility. But one thing was certain: Esteban would never be able to claim any conquest as his own in a Spanish law court. At times the viceroy must have wondered whether this marvelous African was perhaps as dangerous a choice as Castillo might have been. Esteban was an alien child educated by dislocation, a man who had learned the lessons of violence and the whims of the Fates. He was the supplicant of many gods and high priest of a few. In truth, Esteban himself was perhaps far from certain how he would behave when faced with the Indian world once more. His identity was always subject to change and the vagaries of destiny; he was a postmodern man in an early modern world, an inhabitant of the global village before its foundations had been dug. Even now, Esteban was coming to terms with yet another unique experience, the surreal world of Mexico City.

He now found himself in an extraordinary position, almost certainly unique in the African experience of the early Spanish-American Empire. He was still a slave, but he was also a celebrity, almost like a Roman gladiator, part actor, part mercenary. He was the viceroy's houseguest and always in demand at the dinner tables of the most powerful men in Mexico. He must at times have felt as though he had becoming a living version of the Indian legend of *El Dorado*, the "gilded man" who was painted with gold every morning by his servants. He had something of a living myth about him.

One would-be biographer has suggested Esteban now threw himself wholeheartedly into the "good life" of Mexico City. "A lover of women," he made the most of this opportunity and "gave free rein" to his desires. "Like all those of his race, he wore brightly colored clothes when he sauntered through the city streets," all the better to play the part of Don Juan. "For it seems he had a great weakness for seducing young Indian girls who were no doubt attracted by such an exotic figure."

The racial stereotyping in this commentary overshadows an otherwise plausible portrait, for showy attire, amiable drinking, and casual whoring were the regular pastimes in Mexico City of many a leisured soldier who had won his spurs and his gold and who waited unhurriedly for the next call to arms. Such are the ways of men on the frontiers of empire. Esteban was an exception among the African population of Mexico in the 1530s, but he was a celebrated adventurer perhaps quite happy to share the easy life of his Spanish peers. Equally, we would do well not to overemphasize the importance of the two brief Spanish reports of his promiscuous ways.

THERE HAS LONG persisted a romanticized picture of the lives led by the first black slaves brought across the Atlantic by the Spaniards. These pioneer African-Americans were treated more as domestic servants than as slaves and were protected by medieval Spanish laws—so the story goes. This inaccurate image is useful because it clearly contrasts the conditions experienced by men like Esteban with the industrial scale of the southern and Caribbean plantations, which we usually associate with slavery in America.

But that romantic notion of some happy harmony between servant and master is a fiction. From the outset, the Spanish colonies in the New World suffered from a chronic shortage of labor. The Spaniards soon discovered that the Indians did not adapt well to the politics and practicalities of peasantry and serfdom. Their world had been free of feudal suffrage and universal villeiny and they made unproductive serfs who became depressed and indolent in the face of forced labor. Worse, their numbers were falling alarmingly as the disruption to their society led to a reduced birthrate and Old World diseases decimated those who were born.

The problem was exacerbated by the fact that the Spaniards who sailed for the New World arrived with haughty pretensions to nobility and they refused to work with their hands and would not till the land. Spain found a solution in 1501, when it was decreed that Christian blacks, known as *ladinos,* or "latinized" Africans, should be sent to the colonies to provide the necessary manual labor for farming and mining.

Year after year, there was fierce discussion of this policy. There were those on the one hand who claimed that these *ladinos* were well suited to colonial life because they had been latinized and taught European, Christian ways. But others believed that precisely because *ladino* Africans had been "civilized," they had learned to respect themselves as rational men and therefore abhorred their slavery and tended to rebel against their condition, much as a Spaniard might if he were enslaved. Astonishingly, it is an argument that survived in respected historical works well into the twentieth century. There were many who argued, on these grounds, that it would be better to import slaves directly from Africa because they had no experience of European society and culture. Such slaves were known as *bozales*.

The reality was more pragmatic. The colonists needed manpower, and the most productive laborers were Africans. As the colonies grew, there were not enough *ladinos* to go around. The argument that *bozales* were more docile and the *ladinos* more troublesome was largely a convenient way to combat the religious moralists who were worried that slaves brought straight out of Africa might corrupt the Indians with their pagan religions.

In 1518, on the eve of Cortés's conquest of Mexico, that moralizing religious debate about *ladinos* and *bozales* was sidelined by Charles V. In order to raise funds for his own political campaigns, he sold a license for the export of 4,000 African slaves to the Americas. It was the first of many issued to raise funds for the Spanish Crown.

Africans quickly became as much a feature of Spanish colonial life as the Spaniards themselves. Some served their masters as pampered popinjays, manikins for the display of private wealth, to be sure, and many were certainly domestic servants. But every mule train that traversed the Mexican landscape carrying trade goods seems to have had its black teamsters. Thousands of blacks were employed as overseers in textile mills or as stewards on the conquistadors' great estates, slaves whose task was to force labor and tribute out of the subject Indians. Many others labored themselves on the land or in the mills. The lot of Africans in the Spanish New World was diverse.

Slowly, the number of free blacks grew and enough achieved sufficient prosperity for the Spanish Crown to determine, in 1574, that they should be accorded the civic privilege of becoming taxpayers. A royal decree explained that the Crown had heard that because of the great natural wealth of the Indies many black and mulatto slaves were now free and relatively well off. As they were now royal subjects, just like the Indians, these free blacks should now be expected to pay an annual tribute of a silver mark. Clearly, a peacefully settled population of free blacks was expanding in rural Mexico. But, however peaceful, they proved to be uncooperative taxpayers who steadfastly resisted attempts to extract the tribute now due to the Crown. Perhaps it is more surprising to learn that enslaved Africans also managed to own property, although the Crown tried, unsuccessfully, to prevent them from doing so, and also failed to tax them on it.

Africans frequently appear in legal documents as the agents of their Spanish masters in the criminal exploitation of the Indians. They forced their way into Indian homes, stole their wares from them as they prepared to go to market, and organized press-gangs to round up Indian men when there was hard work to be done. No doubt encouraged by their experience of such exploitation in the service of the Spaniards, it seems that many Africans oppressed Indians from time to time for their own benefit. In 1541, the Spanish authorities in Peru complained that black slaves were out of control and went about stealing all that they could from the Indians and even "retained many male and female Indians as servants," all of which was quite "ruinous." Irony, like beauty, is in the eye of the beholder.

The Spanish government reacted to these problems with the usual recourse to grim punishments and by authorizing the Indian authorities to arrest Africans who marauded among the Indians. But they also resorted to a principle of local apartheid that tried to prevent blacks from living in Indian communities at all.

In practice, attempts to separate Indians and Africans proved futile, not least because many more African men than women were exported to the New World. And so, just as Spaniards took Indian women as wives, mistresses, lovers, and concubines, so too did African slaves.

The Spaniards were horrified by any union between Africans and Native Americans and by the cross-fertilization of the two cultures and societies. There were many deep-seated psychological reasons for their horror, which were often manifested as a fundamentalist fear of African and Indian religions. The religious authorities were especially concerned because African and Indian *curanderos*, medicine men and women, adopted rituals and practices from one another and then incorporated aspects of Christianity as well. It was a richly heterodox alchemy that frightened the Church and fueled the zeal of the Inquisition.

Esteban must have been very worried, for he and his companions had survived their long ordeal among the Indians of the north thanks to his ability to combine African and American animism with Christian spirituality. Like his aristocratic Spanish companions, Esteban had every reason to fear the Inquisition.

But while the religious authorities were concerned about such heterodoxy, all Spaniards in Mexico were utterly terrified by the prospect of rebellion. They were not simply paranoid about a slave rebellion in the way that slave owners have been throughout history: in Mexico those fears were exacerbated by the possibility that the African slaves might unite with the free Indian population. There is every reason to believe that paranoia was well founded. Slaves and Indians had every reason to rebel, and they far outnumbered the Spaniards.

In the summer of 1537 that terror exploded into violence, and Esteban saw firsthand the extreme extent of man's unbridled capacity for cruelty to his fellow man.

BY 1537, THERE were as many Africans as Spaniards living in the Spanish district, the *traza*, at the heart of Mexico City. But they were a people apart. The oppression of African slaves on the grounds of geography and skin color led members of African tribes and nations who had treated each other as different and even as enemies when in Africa to regard themselves as unified by Spanish racial oppression. They began to experience a sense of a new and collective identity as Africans that was based on slavery, that was utterly colonial, and which led to a sense of themselves that was uniquely American.

News had arrived in Mexico, in 1535, of a fundamental change in the human rights extended to Africans and Indians by the Spanish Crown. The Crown had decreed that there was too much delay in the sentencing of serious criminals and revoked the long-standing right of all *negros*, freedmen as well as slaves, and Indians to appeal to the high court in Spain in cases where they had been sentenced to death or to corporal punishments such as having limbs or testicles cut off. The latter was a not infrequent punishment meted out to runaway slaves. This must have seemed like a sensible and pragmatic move to prevent endless dissembling by convicts, but in practice it handed too much power to the colonial officials, who were also the slave owners. Far from being an administratively expedient move, it aggravated Africans' antagonism toward the ruling elite. Crucially, it alienated free blacks by sending a clear message that they were still considered "almost slaves."

In such a climate, the Spaniards' paranoia escalated. On the night of September 24, 1537, after a long, stifling summer, Mendoza was working late on the usual business of the colony. An African turncoat, probably a free man who felt himself more conquistador than *negro*, came to Mendoza with the news that the Africans had chosen a "king" and were plotting to rise up and seize the land. "Because my informer was black himself," Mendoza reported to his sovereign, "I did not believe him." But because he had also told the viceroy that the rebellion would involve the Indians, Mendoza "secretly looked into the truth of the matter" by sending some of his most trusted men to spy on the Indians "under cover of darkness."

Mendoza reported that his spies overheard rash talk, no more than "a sign" of a rebellion, but it was enough to frighten him. He immediately ordered the arrest of the African "king," along with the other leading rebels, and then sent out the alarm to all the mining and farming communities. The die was cast and retribution would now be inevitable. "The blacks who were arrested confessed the truth. They had planned to seize the land," Mendoza succinctly reported.

Years later, in 1612, an Indian nobleman called Chimalpahin, recorded details of another plot in his diary, writing that "the blacks of Mexico City had planned to rise up and kill their Spanish masters," while the Spaniards were distracted by their Easter processions. "On Maundy Thursday," as they paraded through the streets, the blacks would "turn them into dead men," or so the witnesses at the trial testified. "This caused such fear among the Spaniards that the Easter processions were canceled."

These African rebels of 1612 had dreamed of seizing control of Mexico themselves. They had chosen a king and queen and appointed a council of ministers and officers of the government. According to Chimalpahin, the Africans planned to subjugate the Indians, forcing them to pay tribute and serve them according to the Spanish model. They planned to murder all the male Spaniards: every man, boy, and babe-in-arms would be killed, except for a handful of priests who were to be castrated. The Spanish male bloodline was to be eliminated from Mexico.

Chimalpahin's imagination perhaps ran away with itself as he expanded on the detail of this plot. All the old and middle-aged Spanish women were to be murdered too. Even most of the young *señoritas* would not be spared. Only those who were very pale-skinned or very pretty were to survive so that they could to be given to the African men as concubines. He also recorded that if these girls then gave birth to boys who were not obviously black or *mulatto*, then the fathers would be forced kill their own children. Pale-skinned daughters, however, would be saved for further breeding. This image of ferocious, vicious revenge exemplifies the terrible fears that paranoia can coax out of brutal oppressors when faced with rebellion by the oppressed. The terror is reduced to the primordial symbols of sexual impotence and the violation of defenseless womenfolk. It is a visceral terror that quickly incites savagery among the fearful.

During the night of September 24, 1537, the Spanish population of Mexico City reacted quickly. Mendoza mustered the conquistadors, 620 cavalry and countless infantrymen. The rebellion, real or imagined, was swiftly and brutally suppressed by this army of frightened, isolated Spaniards.

The convicted rebels were hanged and then quartered. Chimalpahin provides a chilling description of the retribution that followed the similar plot of 1612, when thirty-five black rebels were executed. The executioner, a mixed race man called Cristóbal, hanged the blacks with the help of his son. Eight new scaffolds were constructed for this purpose. Meanwhile, the victims were paraded through the streets on horseback, their upper bodies stripped bare, no doubt showing the marks of torture. At a quarter past ten in the morning, father and son began to string up the convicted rebels in the courtyard of the palace. "All died in agony, but they confessed," and, now at peace with God, they left this life "crying out for forgiveness to their Savior, Our Lord Jesus Christ." By one o'clock, Cristóbal and son had finished their work. The following day, they cut down the cadavers and the magistrates of the city ordered that they be drawn and cut in half.

The remains were to be displayed about the town as a warning to others. But there was heated disagreement in the Audiencia. The more measured burghers argued that "it would be far from a good thing for all the dead to be quartered and hung up and left to rot in the main streets of the city, because their putrid stink would cause such disease as would be quickly carried about the town and it would make everyone sick."

Twenty-nine of the corpses were beheaded and the heads alone were stuck up on the gallows. Their bodies were handed over to the African population of the city, who were allowed to bury what remained of their dead with the help of a number of priests. The other six bodies were quartered and the pieces were hung up on the main roads into town, the usual treatment of the bodies of the most feared criminals during the seventeenth century. Even in death they were reviled.

Following the suspected plot of 1537 and its brutal suppression, when the bloodletting was done, the Spaniards' paranoia increased. An angry colonial official wrote to the Crown, expressing the sense of urgency and fear that had swept Mexico. He claimed to have repeatedly pressed Mendoza to improve the city defenses and was aghast that administrative concerns were impeding progress. It would be better to get on with it, the panicked patrician cried, "for God would not wish some rebellion or uprising to succeed," such as "the revolt which nearly took place only days ago" and which was "arranged by the treacherous *negros.*" For it would be impossible to "save us all," because there was no other city or town "so exposed as this." Spanish Mexico City was in the hands of its enemies, he claimed, surrounded by water and Indian settlements. The terror this correspondent felt is palpable from his hurried, irrational prose.

By the middle of October 1537, Esteban can have had little doubt that, if at all possible, he should ensure his future lay away from Mexico City. He perhaps discussed his options with Juan Garrido, who was now making up his mind to go to Spain. But Garrido was a free man, and Esteban remained a slave—one mistake could cost him his life. The most promising opportunity for escape was obviously the coming expedition to the Seven Cities of Gold. He was of paramount importance to Mendoza and Zumárraga, and his views were listened to. His words could move the minds of powerful men. He could influence the makeup of the expedition, and in so doing could perhaps engineer for himself an opportunity to escape and return to a happier life among the Indians.

FROM THE MOMENT the four survivors first arrived in Mexico City, in 1536, Zumárraga had worked tirelessly, using his influential and highly efficient network of Franciscan contacts to lobby royal officials in Spain for authority over any expedition to the north. Mendoza had also fought hard for control of the expedition, but the viceroy was forced to admit defeat when he received a royal decree from Charles V declaring that the king "had been informed that there were religious men in Mexico of virtuous and exemplary life and high purpose who wanted to journey to newly discovered lands not yet conquered by the Spaniards." The king had heard that "they intended to take the Holy Catholic Faith to the heathen natives and thereby do great service to the Lord." "You are to grant them license for this purpose," was the order.

Esteban was central to Zumárraga's plans for a peaceful, religious expedition. He wanted Esteban to go as a guide and to take charge of the Mexican Indians who would accompany two or three Franciscan friars. He appointed his friend and Las Casas's confidant, the French Franciscan Marcos de Niza, as the spiritual leader of the enterprise. Marcos was politically reliable, had a reputation as a good navigator and geographer, and seemed intrepid. He was by now an old hand in the New World and to Zumárraga, he seemed perfect for the mission of peaceful conquest.

On April 4, 1537, the Archbishop had written to a friend in Spain, enclosing Marcos's signed declaration about the atrocities committed by the Spaniards who had conquered Peru. He was in no doubt about the value of this turbulent priest, expounding that Marcos was "a great man of religion, worthy of our Lord," a man "whose virtues are proven and who is very zealous in his faith." Marcos was so honest a man of God, Zumárraga enthused, that "in Peru the friars chose him as their *Custos*," that is, their monastic leader. Marcos had then left Peru for Guatemala, where he wrote to Zumárraga about further Spanish atrocities. Zumárraga wrote back, asking him to come at once to Mexico.

To understand just how deeply Zumárraga trusted Marcos de Niza, however, it is necessary to look briefly forward almost a decade, to 1546, when Zumárraga lay sick, aware that death might snatch his soul from the world at any time and leave his earthly body to the worms. "My beloved brother," he wrote to Marcos, "I have received your letter and the bed you sent me." But "the cold wakes me," he explained without complaint. "It is a good thing to do penitence now so as not to leave it all for the next life, seeing as this earthly life will not last long."

By 1546, Marcos had retired to the beautiful, bucolic suburb of Xochimilco, which remains today a garden paradise offering heavenly respite from the pollution of Mexico City. Colorfully painted boats are gently punted along the canals which sit, grid-like, embracing the last "floating" fields, the *champas* of old Mexico. On Sundays, middle-class Mexicans flock here to picnic on these barges, while they are serenaded by boatloads of *mariaches* with their broad-brimmed hats, their fiddles and guitars, and their doleful songs of love and romance. But in the early morning, it remains a tranquil place of contemplation, where the cares of the world are washed from the mind by the gentle lapping of the waters against the muddy banks.

Marcos replied to Zumárraga's lament in humble, beseeching

tones, for he was preoccupied with his own material comfort and his own ill health. Zumárraga, he wrote, would be well aware of how Marcos had suffered when he left more temperate climes and knew that was why he had been sent to convalesce at Xochimilco. "As an orphan, with neither father nor mother to turn to for help," he only had Zumárraga, he said. Marcos begged his friend to send him a "little" wine, which he needed for medicinal reasons, "as he was very wan and pale."

Having stated his need for alcohol, he quickly moved on to an explanation of the logistics required for the wine to reach him. There is a strong sense that Marcos de Niza was a drunkard, perhaps an alcoholic, a devotee of Bacchus as well as of his Christian God. Zumárraga wrote back to him by return of post, promising that "while I live and for the duration of your illness, I will send you an *arroba* of wine each month." Marcos no doubt prayed all the harder for the ailing Archbishop's longevity.

Before his alcoholic retirement, Marcos de Niza had lied about what he saw during his expedition to the Seven Cities. In 1539, he had given wildly exaggerated reports about the north, elaborating on his story for almost any audience, eulogizing his promised land from the pulpit and in the barbers' shops, talking of a place where the women wore jewelry of precious metals and the men wore belts of gold. Many were willing to open their wallets; for the price of a few drinks they could cock their eager ears and listen to the latest installment of his adventures.

In 1539, soon after Marcos's return, a Spanish monk in Mexico had written glowingly to the prior of a monastery in Spain that the place discovered by Marcos was heavily populated and had a well-ordered society. There were walled cities with great houses. The people wore silken clothes and shoes. "I will not write about the great wealth of the place, for it is so rich that it will seem incredible to you," the Mexican monk reported to his superior, but then he wrote about it anyway. "The temples are covered in precious stones, I think Marcos said they were emeralds, and the hinterland is home to elephants and camels."

Even the shrewd Zumárraga was taken in and wrote enthusiastically to his nephew in Spain, telling him that Marcos had discovered an even greater land than Mexico. It was 400 leagues beyond the country conquered by Guzmán, near California, a place of large cities and plains where camels and dromedaries roam while quail fill the skies. The whole of Mexico, he said, wanted the chance to join the next campaign.

But Marcos's verbal accounts were a tour de force of fantasy. In reality, Esteban had abandoned him and he had found nothing.

This hindsight about Marcos's exaggerated storytelling and eventual disgrace is necessary to understand the strength of Zumárraga's misplaced trust in Marcos. Even after Marcos's account of his adventure had been exposed as a fake, the old Archbishop was kind to him. Years before, in 1538, even pragmatic Zumárraga had begun to commit the tempting sin of believing delusional tales that he wanted to believe. He laid plans for Marcos's expedition, but his vision was unrealistic, an impossible dream born of his strong spiritual desire to see religious missions replace the cruel expeditions of the conquistadors. His faith, it seems, had blurred his judgment, for he was proposing that two or three Franciscans, guided by Esteban, and protected by a ragtag army of assorted Aztecs and Indians from northwest Mexico, would be able to evangelize all of North America. Even allowing for their uncertain knowledge of the geography involved, it was a far-fetched idea.

Antonio de Mendoza had risen to the rank of viceroy because of his pragmatism rather than his zeal. He no doubt realized that Marcos's mission would almost certainly turn out to be little more than a fantastical charade. Without trespassing too obviously on Zumárraga's royal grant, Mendoza began to arrange matters so that he could have as much influence as possible over the expedition. Most of all, Mendoza seems to have realized that while the expedition would conquer nothing, Esteban was clearly a very able explorer and communicator. The viceroy was determined to ensure that Esteban was given as much practical support as possible, for which it was obviously essential to strengthen the vulnerable military outpost at Culiacán, for a strong northern base camp was central to those plans. There had been worrying reports that the Spanish inhabitants and the garrison at Culiacán were so isolated and impoverished that they were considering abandoning the settlement altogether. The situation there been exacerbated by the arrest of Nuño de Guzmán, leaving a power vacuum in New Galicia. Mendoza now addressed that problem, appointing his protégé and close confidant, the young Francisco Vázquez de Coronado, as the governor of New Galicia. Coronado was to accompany Marcos and Esteban as far as Culiacán, with instructions to strengthen the settlement and convince the inhabitants to remain by means of a judicious mixture of bribery and threats.

Coronado, Marcos, and Esteban set out from Mexico City for New Galicia in the fall of 1538, heading back along the main highway which Esteban had traveled in the other direction two years before.

19

FRANCISCO VÁZQUEZ DE CORONADO

1539

FRANCISCO VÁZQUEZ DE Coronado was full of the dynamism of youthful power, brimming with the vibrancy of Mendoza's favor and a strong sense of his own aristocracy. It was rumored that his fabulously wealthy father-in-law had been the illegitimate child of King Ferdinand the Catholic himself. As governor of New Galicia, Coronado was now a powerful man in his own right.

In late 1538, Coronado set out from Mexico City with responsibility for all of Mendoza's political ambitions in the north. Esteban and Marcos de Niza traveled with him, along with another Franciscan, Brother Honorato, a young friar eager to proselytize the heathen of Arizona and New Mexico. Despite his zeal, Zumárraga had been enough of a pragmatist to understand that the help of Mendoza and Coronado would be essential if his expedition to the Seven Cities was to succeed. He had dictated the religious purpose of that enterprise and had appointed the personnel, but the logistic and military support needed by Marcos de Niza and Esteban was still in the hands of the secular authorities.

Coronado no doubt quickly realized the friars were unrealistic dreamers and that Mendoza's secretly whispered doubts about Marcos's ability to run an expedition were well founded. The Frenchman was a an able orator, but a raconteur rather than a leader. His love of food and wine undermined the gravitas necessary to the command of other men. He evidently had no idea of how to organize his mission. By contrast, Esteban was clearly a charismatic figure who quickly gained the trust of Indians and the respect of Spaniards. But he was an unfathomable cultural anomaly, which made Coronado suspicious and reluctant to trust him too far. The young commander had plenty to worry about.

Coronado became more and more concerned about his own mission during their long journey along the slow road to New Galicia. As they proceeded, they encountered party after party of shackled Indians, forty or fifty at a time, strung together like animals on their way to market. It was Coronado's business to put an end to that illegal trade in slaves, but he could readily see what a challenge that might prove to be, for the sale of those slaves was obviously the economic lifeblood of the province he had been sent to govern. He also saw miserable free Indians, who traveled to and fro between New Galicia and Mexico City. They were *tamames*, human beasts of burden, porters who carried merchandise about the colonies. They seemed broken physically and spiritually by their labors, perhaps even more so than the slaves. Coronado knew only too well that one of the few ways left to the residents of New Galicia to legally profit from their Indian serfs was to rent them out as workmen in Mexico City and then force them to return home heavily laden with goods and merchandise. But that only helped to make the farmland in the province unproductive.

At Guadalajara, Coronado met with an attorney representing the residents of Culiacán who was on his way to Mexico City with especially worrying news. The attorney was gravely concerned and reported that the settlers were preparing to withdraw from the colony in the face of terrifying raids led by an Indian rebel called Ayapín. He even claimed that the provincial capital at Compostela was now under threat and urged Coronado to take swift and decisive action before the whole colony was lost.

The picture was repeated all across New Galicia. Coronado wrote to Charles V, explaining that "Your Majesty should know that most of the Indians in this province are in revolt, for some of them have never been conquered, and others have rebelled against our Spanish rule." In fact, he reported, the only Indians who were at peace with the Spaniards had been commandeered by Guzmán and his cronies for their own estates and private purposes. Now very concerned about the viability of Marcos's expedition, Coronado ordered his army to hurry toward Culiacán.

When they reached the tiny roadside settlement of Tonolá, Coronado decided that they were sufficiently far from the prying eyes and ears of Mendoza's enemies and Zumárraga's rivals to hand over Marcos de Niza's official instructions for his mission. "Exhort and encourage the Spaniards of Culiacán to treat their Indians well," Mendoza ordered the Franciscan. "Give them my guarantee that if they do so, I will favor and reward them." At the same time, "you will make the Indians comprehend that I have sent you in the name of His Majesty to ensure that they are well treated." Tell them that my heart is "heavy because of the injustice and evil they have suffered and that from now on they will be well treated and their persecutors will be punished."

Mendoza's clarity of vision is admirable. He saw that until the Indians were at peace, the province could not be further colonized. But equally, he knew that a successful conquest of the north depended on a robust colony at Culiacán. He had therefore ensured that Marcos and Coronado had the necessary incentives for bringing about peace between the Indians and the Spanish settlers. Mendoza had given Coronado sufficient funds and the necessary authority to try to ensure the survival of the Spanish settlement at Culiacán, at the very least for as long as it might serve Esteban and Marcos as a base from which to explore in the north.

Mendoza ordered Coronado to promise grants of land and other benefits to the settlers at Culiacán, so long as they fell into line with imperial policy. But if they failed in their duty, then they would be punished and all their prior hardships would go unrewarded. Marcos was told to preach the morality of the arrangement, giving a strong spiritual dimension to those incentives. Material and spiritual wealth went hand in glove in the minds of the most successful conquistadors.

Mendoza also ordered Marcos to report on Coronado's handling of the problems at Culiacán. As was usual in the trustless world created by the suspicious, paranoid administration of the Spanish Empire, these two colleagues who had been sent to work together on the edge of the known world were now placed in the unenviable position of knowing that they were spying on each other. Such was the imperial way.

Mendoza's instructions then turned to the main purpose of Marcos's mission, the search for the seven rich cities of the north. He ordered Esteban and Marcos to gather all the Indians from the region they could muster, especially the group of people from farther north who had come with Dorantes, Cabeza de Vaca, Castillo, and Esteban in 1536, and who had settled just north of Culiacán. They were to persuade those Indian migrants to accompany them as guides and to build confidence among the population of the lands through which they planned to travel. When Marcos reached the Seven Cities of Gold, he was instructed to formally claim them for the Spanish Crown in the name of Mendoza and the Church.

Concise, clear orders about Esteban's role were included in these instructions. Should Marcos find a route northward, then he was to take Esteban with him as a guide "whom," Mendoza wrote, "I order to obey you in everything." Esteban was to follow all of Marcos's orders absolutely and without question, as though those orders had come from the viceroy himself. Moreover, Mendoza added that should Esteban fail to do so, he would incur the "usual penalties and punishments meted out to individuals who do not obey those who are empowered to give them orders by His Majesty's decree." The deliberate ambiguity of that threat was not lost on Esteban. Disobedient slaves were whipped and beaten. Insurrection was punished by death. But great conquistadors like Cortés had been gloriously rewarded in spite of failing to obey their immediate superiors. The choice lay with the individual.

When they arrived at Compostela, Coronado discovered a town that barely existed, as Esteban and others had no doubt warned him. In theory, it was meant to be the administrative center of thirty large estates. The thirty Spanish landowners and their households were required by law to live in their own houses within Compostela itself, to ensure they were adequately protected against Indian rebellion. But only ten meager dwellings had been constructed, because the residents were reluctant to live so far away from their estates. They complained that without proper supervision the Indians became lazy, worthless workers or, worse still, began to rebel. The landowners proposed moving Compostela nearer to their estates so that they would have more control over the management of their affairs. Coronado saw no reason to obstruct them, and the process of moving the provincial capital began.

After a brief stop at Compostela, Coronado, Marcos, and Esteban set out on the difficult and dangerous route to Culiacán, 100 leagues or more of steep and craggy roads that wound through mountain ridges and deep gorges. In places the route was deluged by overflowing rivers; elsewhere the precarious road had collapsed into the the deep canyons past which it led. Their Indian laborers worked hard to clear the thickly overgrown road, but it was desperately slow going. And all the while, the sun burned down and the whole party was plagued by the monotonous whine of vicious mosquitoes and other unwelcome, venomous creatures.

Coronado knew that the four survivors had managed to temporarily pacify the Indians at Culiacán on their way through the settlement in 1536, and he had heard that Esteban had played a central role in doing so. Therefore, as they neared the end of their arduous journey to Culiacán, Coronado presumably questioned Esteban about the outpost and the nearby Indian town. Esteban must have described "a fertile and abundant land, with all kinds of foodstuffs," as Spanish commentators had done. He may have explained how he had seen that the Indians would be willing farmers, were it not for for the culture of exploitation and slave raiding, which undermined the true potential of the colony by frightening away the labor force.

When they finally reached Culiacán, the Spanish garrison's preparations for abandoning the settlement were far advanced. The inhabitants were already gathered in readiness to begin their retreat to Guadalajara. In letters to Mendoza and Charles V, Coronado reported that the culture of slaving had not only driven away the Indian workforce but led to Indian uprisings and raids, which were a constant threat to the Spanish colonists. The settlers hardly dared to leave the military camp and could find no way to make a living. In fact, they had now given up all hope of ever doing so. There could be no clearer condemnation of Guzmán's government than the failure to harness the natural resources of this green and pleasant land.

The Indians had fared little better than their would-be masters. To escape the slave raids, they had fled into the mountains, where they were unable to plant their crops or maintain their dwellings. As Coronado reported to his sovereign, it soon became clear to him that the Indian uprisings were a direct reaction to the bad treatment of the Indians by the Spaniards.

Esteban immediately recognized that sorry state of affairs. Almost nothing seemed to have changed from what he had seen during the spring of 1536, when the four survivors had first encountered Alcaraz and Cerbreros in northern Sinaloa. But Esteban had personally set about righting that situation, alongside his Spanish companions and with the support of Melchior Díaz. Had they not brokered a solid peace? he must have asked himself. Had not Melchior Díaz demonstrated a determination to maintain that peace? Again, he was forced to recognize the untrustworthy nature of the men who ruled in that imperial frontier world.

Coronado immediately set about putting things right. His first action was to distribute to the residents of Culiacán money and goods, which Mendoza had sent with him expressly for that purpose. He then confiscated much of the profitable lands held by Guzmán as part of his personal estates and reassigned them to other settlers. Politically, this was a decisive move, one that simultaneously bought the loyalty of Guzmán's former followers and ensured Guzmán would meet strong resistance in the unlikely event that he was acquitted of his crimes and returned to New Galicia. Marcos preached a sermon that sanctified the settlers' covetousness in order to allay any residual sense of treachery.

Now that Coronado had ensured that the colonists would remain at Culiacán, he turned to Esteban to help pacify the Indians of the surrounding area. Some had had already recognized Esteban, and he was soon able to bring his influence and experience to bear. Little by little, he persuaded the rebels who had fled to come down out of their mountain redoubts and abandon their hideouts in the bush. They slowly returned to their fields and settlements and again pledged their allegiance to the Catholic God and the Spanish Crown. The valleys were again filled with the sounds of agriculture as the cautious Indian farmers reluctantly accepted Esteban's assurances that Coronado was their friend and returned to their fields.

Coronado played his part and was mostly merciful with these Indians, although at least one scapegoat was needed to remind any future rebels that the wrath of a Spaniard could be terrible. Coronado wrote to Charles V to describe his long and difficult pursuit of the much feared Ayapín, explaining how he had captured the rebel leader and put him on trial. Coronado had acted as prosecutor, judge, and jury, and his summary justice found Ayapín guilty of rebellion and sentenced him to death. Ayapín was quartered publicly in the Indian settlement of Culiacán and Esteban was yet again reminded that the material wealth of Mexico City and the comforts of the Spanish empire he served were bought at a very high price.

Coronado now freed many slaves who were natives of the region of the Petatlán River, which lay fifty or sixty leagues north of Culiacán, in the wild, unconquered territories through which Marcos, Esteban, and Brother Honorato would soon be traveling. He sent these Indians back to Petatlán with a clear message for their countrymen: there was no reason to fear the Spaniards, because Charles V had decreed that his subjects should in no way harm them. He had ordered that all Spaniards were to cease waging war on the Indians; nor were they to take captives and enslave them.

The Indians at Petatlán were amazed when they saw their brothers march back into town, free men again, bringing an unlikely story of Spanish mercy. The elders and wise men debated how to respond to Coronado's message of peace. After twenty days of discussion and preparations, an Indian delegation set out for Culiacán. According to Coronado, the ambassadors arrived with an entourage of eighty braves, but Mendoza would later claim the real number had been 400.

On meeting with the ambassadors from Petatlán, Coronado confirmed the noble intentions of his sovereign lord, Charles V, and preached the virtues of Christianity to the Indian delegates. With the characteristic crocodile generosity of the conquistador, Coronado gave these Indians handfuls of the usual trinkets handed out by the Spaniards to those they hoped to bribe into subjugation. He gave them clothes, colored glass rosary beads, knives, and little copper hawk bells. With the exception of the knives, these baubles were of negligible value, but the Indians coveted them as something alien and therefore special. Mendoza had supplied Coronado with plenty of these cheap novelties precisely for such a purpose, since it was obviously important to compensate their new allies for Guzmán's crimes.

Marcos and Brother Honorato now began to preach in earnest to these ambassadors from Petatlán. They taught them to make the sign of the cross and together they practiced their prayers. The Indians learned to say the name of the "Good Lord, Jesus Christ" correctly. But at the same time, they were also preparing these influential ambassadors to become an important part of the retinue of their expedition, a Native American escort that would be crucial to securing safe passage through potentially hostile lands.

With solid foundations laid for a peaceful and prosperous colony at Culiacán, Coronado now wrote a quite bizarre letter, dated March 8, 1539, which gives the impression that he had also succumbed to the Spanish obsession with fabulous places. Almost unbelievably, he began it by stating that he was unable "to recall whether I have as yet written to you about the reports I have had of Topira." But these were not the kind of reports anyone was likely to forget. Topira was, he explained, a very populous region lying between two rivers, with more than fifty settlements. He had been told that beyond these, deeper into the territory, there was another land with an abundance of food: maize, *frijoles*, chilies, melons, squash, and turkeys. What is more, he had learned that the people there wore silver armor embossed with the images of various beasts. They wore gold and emeralds and other precious stones set in beautifully crafted jewelry, and their rulers ate off plates of silver and gold. Even their houses were clad with gold and silver!

Coronado expressed his intention to explore deep into the rough and rugged sierras in search of this startling wealth, as soon as supplies reached him from Mexico. With him he would take 150 cavalrymen, 200 infantrymen, the usual array of black Africans and Indian auxiliaries, and as much food as he could arrange—including an army of pigs that would march alongside the men for whom they would eventually be dinner. It was a huge army to take into bleakly hostile terrain, and what he may have really been going in search of is a matter for conjecture.

It would be easy to dismiss this report of Topira as yet another example of the Spaniards' ability to persuade Indians to tell them whatever story seemed sweetest to Spanish ears. It certainly seems likely that Coronado felt there was sufficient cause to investigate, and presumably he turned to Esteban for help in trying to clarify the situation. But Esteban must have been puzzled and suspicious, for he could see that Coronado was a pragmatist, not a fool ready to believe a fantasy. He must have wondered why Coronado seemed to be bent on pursuing so obvious a fiction. What, Esteban must have wondered, was the young governor of New Galicia up to?

The answer became obvious soon enough. Coronado now explained to Marcos and Esteban that he intended to march through the mountains and drop down onto the coastal plain again at the Indian settlement in the far north which the four survivors had called Corazones, the "Town of Hearts." The exaggerated reports of the wealthy land of Topira seem to have been a ruse used by Coronado to justify outflanking Marcos's supposedly religious expedition with a simultaneous military invasion of Indian country. In reality, the topography of the Sierra Madre would have made it impossible for Coronado to reach Corazones through the mountains, but he had very little knowledge of the terrain when he set out from Culiacán in 1539 pretending to look for Topira.

ESTEBAN WAS THE only non-Indian at Culiacán who had seen the rich arable land of Sonora, the deserts to the north, and the fertile valley of the Rio Grande. Only Esteban had been high up the Rio Grande, where the Indians told tales of nearby towns with many-storied houses. Only Esteban knew that those tales had seemed like fine stories about real towns to him and to his companions when they heard them. Back then, there had been rich promise in the Indians' images of wealthy cities that buzzed with the industry of agriculture and the politics of urban life. But Esteban now understood that the image was relative, that the pueblos were little more than large villages. He had also realized that through some dark, unfathomable imaginative process, the Spaniards of Mexico City had come to believe in the reality of the Seven Cities of Gold. Esteban knew only too well that their goal was nothing more than an illusion. But even Esteban must have been influenced by the peculiar excitement of their improbable venture and the hope it offered of a different life among reverential Indians rather than disdainful Spaniards, a life he no doubt began to long for.

Esteban, Marcos, and Honorato now prepared their party. For twelve days all was hustle and bustle in Culiacán as they sent messengers ahead along the road and busied themselves with packing the provisions they would need. The trinkets and novelties with which Marcos hoped to buy the love and affection of those he met were carefully wrapped in bundles of woven grass. The old-timers at Culiacán no doubt teased the friars with stories of hardship and sexual conquest in Indian country. But few can have teased Esteban. He was no doubt viewed with awed suspicion, private wonder, and superstition. Dressed like an Indian shaman, Esteban looked like a man who could call upon the darkest of spirits; he looked like a man who might be in league with the devil. But everyone knew that success or failure would depend on him.

Esteban, the irregular army of Indians, and the two Franciscans set out from Culiacán on March 7, 1539, and headed for Petatlán. They traveled the well-used road that for centuries had carried the trade and traffic of prehistory between Mexico and the Pueblo Indians of Arizona and New Mexico. Mendoza's policy of diplomacy and reconciliation had clearly worked: the expedition was greeted by thousands of Indians who appeared from their unseen settlements deep within the dense brush of thorn and mesquite. When night came, these Indian hosts quickly built wigwams from woven mats of reeds or rushes, supported on branches that were swiftly and skill-fullywrenched from the bush. They brought the travelers sweet-scented flowers, food, and drink.

A Spanish chronicler called Pedro de Castañeda later reported rumors of the deeply uneasy relationship that developed between the fearless African adventurer and the fearful Franciscan friars during this journey. Esteban quickly fell into an easy rapport with the Indians, who understood his character and behavior much better than they understood the strange ways of the Christian missionaries. The Indians knew Esteban from his earlier visit, and he knew them. He spoke their language and wore the signs and symbols of a shaman and carried the scepter of a medicine man. The Indians revered Esteban as an agent of a spirit world that promised salvation. Castañeda reported that they offered him rich turquoises and buffalo hides and even their wives and daughters, all of which the uncomprehending Franciscans found indecent and immoral.

Esteban remembered the spectacular stage scenery that Aztec craftsmen in Mexico made for Christian festivals and how it had pleased the friars and churchmen. He now persuaded the Indians he met on the road to Petatlán to do something similar. Marcos later described how in places, the inhabitants constructed rudimentary triumphal arches to line the route as a greeting to the two Franciscans. They also shared the little food they had with their visitors. Marcos later commented that the Holy Ghost had guided him through those lands, but it is obvious that Esteban was truly the immanent dove of peace.

As they approached the Indian town of Petatlán, they were again warmly welcomed. The people of Petatlán were skilled archers, excellent hunters who treated their guests to a feast of game. But as Marcos contemplated the rough-and-ready world of these hospitable people, he must have wondered whether these men in tanned deerskins and their bare-breasted women, "dark and ugly," dressed in roughly woven skirts, living in rudimentary houses made of reed mats, were in fact a bad omen for the expedition.

Esteban, by contrast, must again have felt the thrill of traveling as a shaman in Indian country and the unforgettable sense of freedom and uncertainty that comes with continual precipitous movement toward an unknown destination. He felt the power of knowledge as he reassured himself that he alone knew the real truth about the uncertain substance of their goal. Again, he felt the power of realizing that he alone knew the ways of the Indians.

At Petatlán, Brother Honorato fell sick, perhaps intuitively fearful of the road ahead. After three days, Marcos decided to continue the journey without him. It seems an astonishing decision and Marcos offers no explanation for it. Why did he not stay to help? Why should Marcos abandon his only European companion, a fellow Franciscan who was sick? How could he preach alone to so many Indians? Was Honorato truly ill? Did Marcos really take this decision, or was it, in fact, taken by Esteban?

Whatever the explanation, Esteban and Marcos now forged ahead, moving quickly. The region was abuzz with news of Esteban's return, and they were visited by Indians who came from near and far, among them Seri fishermen from the coastal islands, who wore necklaces of oyster shells. Marcos later reported that when he showed these people a sample pearl, they told him that they had plenty on their islands, but, we know the fanciful friar was always quick to whet the appetite of a covetous audience without regard for the truth.

Esteban now led Marcos across a region of uninhabited lands for four days, until they reached another tribe of Indians who Marcos claimed "were astonished to see me, for they had never heard of Christians." But, he said, he was enthusiastically received and the Indians called him "Sayota," which in their language meant "a man from the heavens." There was also abundant food, and Marcos may have felt for the first time like a true evangelist, a preacher of God's gospel among a welcoming pagan people. But, once again, the hand of Esteban is unmistakable in the organization of his reception.

Marcos later claimed to have heard that four or five days' journey inland there was a wide, fertile vale in the foothills of the mountains where a cultured people had plenty of gold and knew about metallurgy. He reported that the highly civilized people of this valley carried golden spatulas with which to wipe the sweat from their brows. But instead of immediately going in search of that fabulous inland paradise, Marcos decided that it would be better to follow the viceroy's instructions exactly, which exhorted him to keep near the coast rather than travel inland. Marcos de Niza's account of the expedition is filled with similarly inexplicable decisions. We can only assume that he invented the fertile inland vale on his return to Mexico City. When Coronado searched for it later, he found nothing but a few miserable huts.

Marcos and Esteban now continued for three days, going due north and in fact getting farther from the sea, to a large settlement called Vacapa. The location of Vacapa has become critical to an important scholarly debate about the reliability of Marcos de Niza's account of the expedition and whether the accusation that he was "liar" is fair. As that account is the principle primary historical source of information about the rest of Esteban's life and the source of the remainder of this history, the credibility of Marcos and his report are of paramount importance here.

For centuries, Marcos de Niza has been maligned as a fraud, with a penchant for make-believe, and not without reason. The sixteenth-century soldier, explorer, and historian Baltasar Obregón summed up his contemporaries' disillusionment with "the deceit and lies with which Fray Marcos de Niza publicized and told his story." He told it "to such effect, that within no time at all, most of the citizens and inhabitants of Mexico City were in a greedy and frenzied hullabaloo," keen to join any future expedition. Cortés also attacked Marcos for being duplicitous, incompetent, and an outright liar. But in the twentieth century, many scholars defended Marcos's reputation and to this day there has been no unanimous verdict convicting him. The essential positions of the prosecution and the defense were laid out in 1930s and 1940s in a debate conducted in the pages of the *New Mexico Historical Review.*

The case for the prosecution was opened by a distinguished Berkeley archaeologist and geographer, Carl Ortwin Sauer, who worked tirelessly for many years in the southwestern United States and northwestern Mexico. Sauer gained very good firsthand knowledge of the lay of the land all along the route taken by Marcos and Esteban, and concluded that Marcos could not have traveled as far north as he claimed, because the times and the places the friar was supposed to have visited simply would not square with the physical distances he was supposed to have traveled. In other words, Marcos could not have covered the ground he claimed to have covered in the time he claimed it had taken him to do so. As Sauer put it, Marcos's report is "convicted of fraud by its own calendar."

Sauer was not alone in this interpretation. The same conclusion was reached by Cleve Hallenbeck, the meteorologist who insisted that Esteban was an Arab or a Berber.

The case for the defense rested on demonstrating that Sauer and Hallenbeck were wrong and that Marcos had indeed reached the Seven Cities as he claimed in his report. The location of Vacapa was crucial to that debate. Marcos's scholarly detractors preferred a more southerly site, whereas his defenders preferred somewhere farther to the north and therefore nearer to his goal.

The defense was given considerable impetus when a history professor called Lansing Bloom noticed that the transcription which Sauer had used in his work did not tally perfectly with the original manuscript kept in Seville. But Sauer then retorted that these discrepancies were not relevant to his argument. In fact, Marcos's report is so vague that it is impossible either to identify his route or to make a calendar of his progress. Every historian who attempts to sort out the mess must inevitably choose to prefer one possible interpretation of Marcos's inexact descriptions over another. But Marcos's guilt can be demonstrated in other ways.

It was Sauer's knowledge of the terrain that led him to judge Marcos's report as untrustworthy, but he also doubted the sensational claims Marcos made about the size, wealth, and potential of the Seven Cities. Sauer, sober scientist that he was, came to the conclusion that "the friar should be remembered not as a discoverer" of New Mexico, "but as one of the most successful publicity agents in our history."

Another investigator, Henry Raup Wagner, dismissed Marcos's account of the route, detail by detail. Significantly, he took a different approach, turning to contemporary accounts of Marcos's character. "The question is largely one of the credibility of the man" himself, "and to determine" "the extent of that credibility, or lack of it, we can only rely on the record of his life and services."

In this battle for "hearts and minds," the defense could call Archbishop Zumárraga as a powerful character witness and point to the fact that Marcos de Niza eventually became a senior provincial in the Mexican Franciscan church before his retirement at Xochimilco.

Above all, perhaps the most eloquent condemnation of Marcos is the testimony of Coronado and the actions of his men. When Marcos returned to Mexico in the autumn of 1539 with his glowing reports about the Seven Cities, Mendoza quickly put Coronado in charge of a great expedition to conquer and settle that promised land. Coronado and his men took Marcos with them, but it soon became clear that the friar had never been as far north as he claimed. As the expedition marched out of northern Mexico and into modern Arizona, Marcos assured Coronado that there was an excellent road leading to the

Seven Cities, which had only one steep grade, of little significance. Instead, they had to follow a terrible trail, struggling along a track which was broken and indistinct and which became ever more dangerous as they climbed into the mountains. Marcos's promises about the road and riches ahead now collapsed one after another, until the furious army turned on the friar. Coronado found himself forced to send Marcos back to Mexico City before he was murdered by the angry soldiers.

Marcos almost certainly never entered the territory of the modern United States during his first expedition, and his report does not withstand proper critical scrutiny. In his defense, some scholars have argued that Mendoza himself persuaded Marcos to affirm legally that he had seen the Seven Cities and that he had claimed them for the Church and the Crown. Sauer argued that Mendoza ensured that the report was heavily embellished in order to encourage Spaniards in Mexico to join Coronado's expedition. "It is time," Sauer wrote, "that the story of the discovery of the Seven Cities by Fray Marcos be classed where it belongs, as a hoax devised in the interests of Mendoza's *Realpolitik*."

There are scholars who disagree. Mendoza and Coronado, they argue, invested so much money of their own that they must have believed in the Cities of Gold. It is impossible, they say, that these men would have mobilized Coronado's great expedition to the north and spent 85,000 gold pesos each if they had themselves helped to concoct Marcos de Niza's fictional account. But the whole history of the Spanish discovery and exploration of America is a story of bold, foolhardy men acting on a whim and a prayer. There was nothing unusual in losing a fortune on expeditions into the unknown. Even Cortés was, at that moment, in the process of almost bankrupting himself by sending ships to Baja California in search of the very same prize. It was an enterprise in which he had been engaged for many years. Mendoza had quite rightly set his own sights on the Great Plains, not the

Seven Cities, and presumably believed there was money to be made out of raising stock there. To Mendoza, Marcos's account was indeed a useful encouragement for the conquistadors he hoped to recruit.

When the evidence is weighed in Probability's balance, we are drawn toward the conclusion that Marcos was indeed a liar and a charlatan. He may have been encouraged by Mendoza and Coronado, but he embraced his role. As a result, it is very difficult to know how much truth or how many lies about Esteban may be contained in Marcos's report. But that report is almost all that we have to work with, and is by far the most significant documentary evidence of Esteban's fate.

20

CHRONICLES OF ESTEBAN'S DEATH

MARCOS DE NIZA states in his report that Esteban left Vacapa on March 23, 1539, Passion Sunday, a week before the start of Holy Week. It was the last time any European saw Esteban, and it is the last moment in Esteban's life for which we have an eyewitness report. Marcos said that he ordered "Esteban Dorantes, the black man, to follow the northern route for fifty or sixty leagues ahead in order to see whether he came across information about the great things that we were searching for." As soon as Esteban heard news of something important, he was to either return to Vacapa in person or else remain where he was and send word to Marcos.

Marcos had devised a simple but highly symbolic form of communication for Esteban's messages to him. If Esteban learned of a "reasonably sized settlement," he was to send to Marcos, by Indian messenger, a small white cross, about the size of a human hand. If he got wind of something of really major importance, he was to send a white cross the size of two hands. Should he hear of somewhere larger and richer than Mexico itself, then he was to send a really large cross. The bigger the material prize, the bigger the symbol of the Savior's suffering for the venality of man. Marcos was truly a man of his time and place, adept at concocting an appetizing potion of moral materialism and sacred idiosyncrasy as a means of receiving messages from his guide.

Esteban set out from Vacapa with his retinue of Indians, while Marcos de Niza settled into a more comfortable and sedentary life at his new base camp. The housing was Spartan, little more than a tent made of woven grass, and the fare was simple—melon, corn, beans, and whatever game the archers brought home for the pot, washed down with some mildly alcoholic drink made from mesquite. But for Marcos de Niza, it was more comfortable than life on the road and it afforded him the opportunity to preach in a settled community as yet untouched by Guzmán's thugs. Marcos was, after all, a man of God, and the treasure he longed for was the conversion of innocent Indians to Christianity. Whatever ill we may speak of him, he was a man of peace.

While the friar busied himself with his spiritual mission, he sent Esteban to deal with the material business of the expedition. The chronicler Pedro de Castañeda, who had ample opportunity to speak with Marcos, later reported that soon after Esteban had left Vacapa, the African "had come to the conclusion that he could gain all the reputation and honor of conquest and discovery for himself." He realized that "should he alone be the man to find these famous settlements with their tall houses and great wealth, then he would be renowned for his bravery and courage. And so," the chronicler recorded, Esteban "proceeded with his followers," in search of the Seven Cities. But Castañeda cannot have known what conclusions Esteban may or may not have drawn about his own mission *after* leaving Vacapa, because—to repeat—no European ever saw him again. Whether the conjecture was invented by Castañeda or Marcos is uncertain, but as is so often the case, the details surrounding Marcos's expedition seem as much fiction as fact.

The greed for wealth and glory that Castañeda or Marcos had attributed to Esteban was the usual motive common to all Spanish conquistadors, but there is every reason to think that Esteban did not share their vision of how such riches might be achieved. He was a slave, and, although he might hope for his freedom, he could easily foresee that any significant material wealth he might find—any riches that took the form of gold, land, and Indians—would belong to Mendoza or Andrés Dorantes. He had also seen how treacherous the Spaniards were and how brutally they treated slaves, Indians, and even each other. The massacre of Africans in Mexico City was still a vivid memory.

Once we strip away the speculation about motive from Castañeda's account, we are left with the simple fact that at some point after leaving Vacapa, Esteban decided to disobey Marcos's order to wait for him fifty or sixty leagues farther along the trail. Instead, Esteban put as much distance as he could between himself and the friar, and it seems likely that he had decided to escape. He had become a *cimarrón*, a runaway slave.

When Esteban set out from Vacapa, he was already almost beyond the reach of the Spaniards, but he was not yet completely safe. Coronado had announced his intention to seek a route through the mountains to Corazones, the "Town of Hearts." Esteban probably guessed that was an impossible quest, knowing that the terrain was almost certainly impassable, but he could not be sure. To be safe, he had to leave Corazones far behind as well. Only after he had journeyed much farther north and reached the sedentary, civilized people of the upper Rio Grande could the eternally wandering Esteban perhaps at last find rest.

But what if Marcos turned tail and fled back to Culiacán with news of Esteban's disobedience? Perhaps the friar would send word to Coronado that his Afrcian guide had deserted, and a posse of Spanish cavalry would be sent to hunt Esteban down? A renegade slave might prove the perfect excuse to send a military force into the disputed domains of northwestern New Spain. Esteban knew that if he was to be quite sure of escaping, he had to persuade Marcos to follow him. He had to lead the friar as far as possible, luring him into following the trail, until he too was a very long way from Culiacán.

Esteban worked his magic, curing the sick and taking part in shamanic ceremonies. He meticulously ensured that his Indian allies would treat Marcos as a god. He convinced the Indians that the friar had been sent from heaven to teach them about the powerful Christian God, told them to show reverence to Marcos, and taught them how to pray to the cross.

At the same time, Esteban thought about how best to lure Marcos de Niza away from Vacapa. He made the obvious move and sent messengers to Marcos, carrying a huge cross, "as tall as man." The specified meaning of this sign was clear: he had found something greater than Mexico.

And then, in a stroke of genius, Esteban appropriated the Indian name for the six Zuni pueblos that were far away on the borders of Arizona and New Mexico and used it to his own advantage. The Zunis today call themselves *A:Shiwi*, Esteban called their lands Shivo:a, which sixteenth-century Spaniards wrote as Çivola or Cibola. By re-christening the Seven Cities of Gold with that Indian name, he made them seem tangible. But he seems to have almost overplayed his hand, for among the messengers Esteban sent to Vacapa was a man who told Marcos that he himself had visited Cíbola. Even Marcos, with his love of exaggeration, later reported that the man "told me such great things about it that I stopped believing him." The man had said that Cíbola was a city of grand houses with doorways decorated with turquoise and precious stones. There is quite possibly some truth to this story, for archaeologists have found ruined Zuni doorways that were once decorated with green stones.

Esteban no doubt had little idea what Zuni itself might really hold in store, but he knew that it was a very long way away. Marcos would have trouble getting there and even more trouble getting back. Esteban's messengers pressed home the urgent plea that Marcos should follow him on the trail to Cíbola.

The friar was slow to react, hampered, perhaps, by the onerous responsibilities of Holy Week. He remained at Vacapa for ten days, until the Monday after Easter; and by the time he reached the proposed rendezvous, Esteban had moved on. That established a pattern— the friar was enticed deeper and deeper into northwest Mexico, while Esteban always remained slightly ahead of Marcos, tantalizingly out of reach. And he made sure that Marcos was always welcomed by crowds of adoring Indians who eagerly listened to his sermons and quickly learned how to pray and make the sign of the cross. They fed him well and looked after him. They told him more and more about the wealth of Cíbola. One tribe of Indians told him that the people of Cíbola "wear long cotton tunics with wide sleeves," which, he deduced, were "like the clothes of Bohemia." They also told him that "beyond the Seven Cities there were three further kingdoms that were richer still." Marcos de Niza took the bait.

Meanwhile, Esteban himself pressed on, his progress perhaps frustratingly slow as hundreds turned out to meet him, bringing their children to be blessed and their sick to be cured. Patiently he attended to his duties as an itinerant medicine man. Esteban, eager to learn more about Zuni, pressed the Indians for information. Many were eager to tell him what he wanted to know. They traded with Zunis, and that was how they had acquired turquoise gemstones which they set into the spectacular nose rings and earrings that they wore. They showed him fine bison hides, their usual payment for working as agricultural laborers at Zuni. The Zunis, they said, were stoical people, but very hospitable and generous to their friends. They gave Esteban the general impression that the Zunis were a good people apt to drive a hard bargain.

At Corazones, the "Town of Hearts," Esteban was greeted by old friends who remembered him from the winter of 1535, when the four survivors had passed that way. It was a beautiful settlement, surrounded by irrigated fields, where running water and gentle breezes cooled the afternoon air, as in an Arab garden. The local rulers dressed in cotton finery and wore elaborate turquoise jewelry. They brought him deer, rabbit, and quail, corn and sweet tortillas. He lingered, on the brink of his final freedom.

The people of Corazones were eager for definite news from the south. They had heard terrifying rumors about alien invaders, but no one had seen any of these strange foreigners yet, they said. Esteban was relieved; clearly Coronado was still far away and had failed in his attempt to cross the mountains. But the Indians were very worried, because their town had been struck by severe epidemics. What, they asked, is really happening?

He told them that an important "white god" would soon arrive to bless them and ensure peace in their land. He told them to send word into the surrounding valleys so that all the people could come and revere this man of heaven. Then, Esteban ordered them to raise up a large white cross, a clear message to Marcos that the Seven Cities of Cíbola were close at hand. He also explained to the people of Corazones that the cross was a symbol of God.

Esteban now left Corazones and turned upstream, marching quickly through the rough gorge above modern Ures and into the upper Sonora valley. He was back in the "Land of Maize."

THE SONORA VALLEY is a narrow strip of verdant paradise, a place of moist meadows set in a bone-dry ocean of inland hills that rise in soaring ridges to the Sierra Madre. The burned gray-green of the highlands intensifies the bucolic green of valley floor, where the gentle rustle of the river runs through plowed fields and woods and hedgerows that might be in England. Here, Esteban could at last relax, turn his attention away from escape, and begin to contemplate his future. He again pressed his hosts for further information about Zuni.

Many Sonorans had gone to Zuni to trade and work as migrant laborers. They told him that there were six Zuni settlements in a beautiful river valley. They were the nearest of many sedentary tribes who lived in towns to the north. The buildings in Zuni, the Sonorans said, were strongly constructed out of stone, wood, and adobe and the people were hunters and farmers, who grew corn, squash, and beans in irrigated fields. But it was a land of fragile fertility, he learned; the people worshipped water, and their rain gods were considered as vital as the sun himself.

The Zunis were a successful trading nation, and trade had made them wealthy. They bought the magnificent hides of the buffalo that were hunted on the Great Plains and tanned by the hunters themselves. They also traded in abundant turquoise that came from another powerful neighboring nation in the east. From the south and the west they received feathers and seashells in exchange for the hides and the precious stones. Again, he was told that although the Zunis were tough, wily merchants who were usually very reserved, they were at heart a hospitable, friendly, peaceful people.

The six Zuni villages and the river valley were dominated by a round, red mesa, a flat hill that they called Dowa Yalanne, the "Corn Mountain." This was a sacred place, he learned from his informants, and it was also a perfect natural fortification where the Zuni took refuge in times of trouble. A bold red castle of natural rock dominating the plain below, like the red crag on which the Moorish sultans of Spain had built their fabulous fortress in Granada.

What might the now fugitive Esteban have thought when he heard these stories of Zuni? Theirs was a land renowned for prosperity and an urban society, a place of high culture with a reputation for friendliness. Beyond their land were other nations and other cities, places perhaps mercifully beyond the reach of Spain. Beyond Cíbola was a boundless world of possibility where a man was free to dream or make his dreams come true, and it beckoned Esteban on. Rested by his stay in the Sonora valley, he now resumed his journey, following the trail north, across the modern international border and into Arizona.

He made his way along a good road up the then fertile and well-populated San Pedro River toward Chichilticale accompanied by 300 people from Sonora, in addition to the retinue of followers who had joined him as he traveled and the Indian soldiers from Culiacán. From Chichilticale, he ascended into the uninhabited Mogollon Rim, the steep, 2,000-foot wall of sandstone and limestone cliffs on the edge of the Colorado Plateau that is named after an eighteenth-century Spanish governor of New Mexico, Juan Mogollón. As the sun dips low in the west, the escarpment glows a golden rich red against the deep green of the ponderosa pine stands. It is a breathtakingly beautiful place and the final natural obstacle on the road to Cíbola. Esteban now climbed the Indian trail that followed a serpentine course between the crags and gullies of those imposing cliffs.

MEANWHILE, AN EXHAUSTED Marcos de Niza arrived in Sonora and was astonished to find so heavily populated a valley, which with its fine irrigation systems seemed to him like some "great market garden." He was so impressed that he declared in his report that up to 300 cavalrymen could be stationed there, with plenty of fodder for the horses and food for the men. His mind was already turning to thoughts of conventional conquest.

While Marcos was thankful for a chance to rest, he was also becoming concerned that Esteban had not waited for him, and as he took stock of his situation, he must have begun to feel isolated and vulnerable. He was a long way from Culiacán, and it was clear to him that Coronado could not have reached Corazones and almost certainly never would. Esteban and the Seven Cities had become alarmingly elusive, and Marcos's thoughts must have turned increasingly to the comforts of monastic life in Mexico City.

He now questioned the Sonorans about Cíbola, anxious to establish how much farther away it might be. Their replies did nothing to allay his fears, for they told him that Cíbola was still many days' march, through a long, populous valley, then up into the mountains, through pine forests, and across a desert. He had every reason to be reluctant to continue his journey, but he was also intrigued by the Indians' constant conversation about Cíbola.

It seemed to Marcos that the people of the Sonora valley talked about nothing else. They described in detail the construction techniques used to build the towering houses of the Seven Cities. When Marcos refused to believe them, they showed him how it was done by building little models with dirt and pebbles. Then Marcos reported that he had met an old man "who was a citizen of Cíbola." He was even more civilized than the people of the Sonora valley, and he told the friar that "Cíbola was a large and populous place, with buildings that were as high as ten stories where the most powerful men met on certain days of the year" to speak their minds. Cíbola, it seemed, had a parliament of sorts, a forum for governmental debate. This old raconteur then managed to convince Marcos that there were neighboring kingdoms, at war with Cíbola, which were richer still. He explained that he himself had been forced to leave Cíbola because he had fallen out with the governor. He asked Marcos to intercede on his behalf.

This Zuni refugee is a puzzling character who immediately disappears from the account, without explanation—an extraordinary lapse, given that Marcos's mission was to discover as much as he could about the Seven Cities. He should have insisted on lengthy testimony from this informant and written that testimony into his report. He should have tried to bring the man back to Culiacán. He should have at least offered some explanation of why he did not make the most of this excellent opportunity. But instead, the old Cíbolan evaporates in Marcos's narrative as quickly as he appeared, like a djin in an Arabian tale. This Cíbolan is perhaps a fictional character, a make-believe witness to the existence of the Seven Golden Cities that Marcos de Niza failed to reach. He seems like a nonexistent citizen of a fantastic place that existed only in the minds of the deluded conquistadors.

Marcos reported that while he was in the Sonora valley, enjoying sumptuous hospitality, he was given the last message that he ever received from Esteban. According to his account, Esteban sent word explaining that he was now in an exuberant mood and was crossing a final unsettled area before reaching Cíbola. In this message, Esteban assured Marcos that "he was now quite certain that everything they had been told by the Indians about their goal must be true because, as yet, the Indians appeared to have told him nothing but the truth. At least, he had not managed to catch them telling any lies." That, at least, is the message that Marcos claimed to have received from Esteban.

But this is obviously fiction, because Esteban had no way of sending such a complicated and detailed message, as Marcos made clear when he described the necessity of communicating using his peculiar system of different-size crosses. While he obviously chose to use crosses as the medium of communication for reasons of religious symbolism, it is more important to note that he needed a simple language based on signs at all, presumably because there was no other practical way for the two men to communicate. Evidently, the Indian messengers could not be relied on to accurately convey the substance of even quite short messages. Equally, the implication is that Esteban was illiterate; and this seems likely, as he appears to have had little opportunity for a formal education. He was almost certainly unable to write, although he had perhaps learned the rudiments of reading. Therefore, Esteban could not have sent a message describing his intellectual reasons for believing the Indian accounts of Cíbola, the kind of message that Marcos claimed to have received from him. Once more, Marcos was inventing a story.

Marcos de Niza then reported to Mendoza and the Crown that on

May 9 he set out himself from the head of the Sonora valley and after four days entered the last unsettled area before Cíbola. This report makes no sense at all, because it fails to take into account the inhabited San Pedro valley, the Gila River, and the spectacular scenery of the Mogollon Rim. In reality, the last uninhabited region before Zuni was much farther than four days' travel from the Sonora valley.

There can be little doubt that from this point onward, Marcos's account of his own geographical progress was invented, almost certainly to secure imperial possession of the Seven Cities. He reported that he eventually viewed Cíbola from a distance, perched on a hilltop. But he said that there was only one city, whereas in reality there were six villages spread out along the valley. He wrote that while he stood on his hilltop "it seemed to me best to call that land the New Kingdom of Saint Francis. And there I built a great cairn of stones with the Indians' help and on top of it I placed a small, slim cross. And I stated that I had put that cross and cairn there in the name of Don Antonio de Mendoza, the Viceroy of New Spain." Under Spanish law, as far as Spaniards were concerned, Marcos had officially discovered Cíbola. The fact that he had never been there was all but irrelevant so long as no one could prove it.

The reality, clearly, is that Marcos de Niza remained in the Sonora valley, trying to gather enough information about Cíbola from which to concoct a plausible account of the place. He also waited eagerly for news of Esteban, and he appears to have had such news soon. But again, it is impossible to satisfactorily sort fact from fiction in Marcos's account of what happened to Esteban when he reached Zuni. Marcos claimed that he had himself toiled for twelve days across desolate terrain, when "the son of one of the Indian lords who was traveling with me arrived." He was one of Esteban's company; his face and body were covered in sweat; he had been battered and bruised and was black-and-blue. He wailed in anguish and agony, to the terror of Marcos's Indian companions.

Marcos reported that this Indian then told him that Esteban had made his camp a day's journey from Cíbola and sent his messengers to the town to announce his arrival—his usual way of introducing himself to a new tribe or nation. The messengers took his shaman's rattle with them. It was his Indian scepter, a symbol of his power, a dried calabash with red and white ribbons, plumed feathers, and jingle bells, which in Esteban's experience gained him respect and reverence. But now, when his ambassadors arrived at Cíbola and handed his precious rattle to the rulers who had come to receive the delegation, one of the lords, presumably one of the bow priests, took the rattle in his hands and, seeing the bells, dashed it violently to the ground.

"Leave!" he ordered. "I know who these people are and they will never enter our town. Instead, we shall kill them all!"

The messengers went back to Esteban with this unhappy news. But Esteban was not perturbed.

"This is nothing to worry about," he declared. "The fact is that the people who show most anger in their greetings always show me the most hospitable welcome in the end."

With that, Esteban continued toward Cíbola. But when he arrived, instead of welcoming him as he expected, the rulers barred his entry and he was forced to wait in a large building outside the town walls. While he was there, the townspeople came and stole everything he had. They took the gaudy Spanish baubles he had brought as currency for trade, the gorgeous turquoises he had been given by the Indians of Sonora and northwest Mexico, and his valuable buffalo hides. That night, Esteban and his entourage waited, apprehensive, denied both food and water.

The noble Indian youth who supposedly gave this account to Marcos then reported that the following morning he woke early and went to the river to quench his thirst. As he looked up from the cool refreshing water, he saw Esteban running for his life, chased by an army of Cíbolan archers "who killed some of his companions," the man explained.

"When I saw all this," the Indian said, "I hid myself well and followed the river upstream until I could make my way to the trail across the desert."

Marcos de Niza then claimed that despite the misgivings of his Indian guides and companions, he now continued along the desert trail toward Cíbola. The fact that Marcos's description of the route he took and the traveling times he gives make no sense when compared with the real geography of southern Arizona must be emphasized at this point, because the next stage in his narrative reinforces the impression that his account of his own journey was fabricated for literary effect and political expediency.

"A day's journey from Cíbola," Marcos stated, "we came across two more Indians who had been a part of Esteban's retinue. These men were covered in blood and scarred with many injuries." The Indians who accompanied Marcos wept bitterly out of pity for these ill-treated men and out of fear for their own safety.

Marcos de Niza calmed the frightened men and persuaded them to tell their story. They told more or less the same tale as the battered Indian prince Marcos had interviewed a few days before. They confirmed that when the messengers reached Cíbola with news of Esteban's arrival and carrying the shaman's rattle, the lord of Cíbola had indeed hurled the sacred object to the ground, and had then spoken these words: "I know what these people are about, for these jingle bells were not forged in our smithies. Tell these people to go back from whence they came, or else not a man among them shall live."

The Cíbolan lord stood shaking with rage, and the saddened messengers returned to Esteban. At first they did not dare to tell him about their reception, but when they did, Esteban rose to the challenge of adversity and spoke to his men, telling them to have no fear.

"I will go there myself, because for all that they have reacted so badly to my messages, they will welcome me warmly."

And so Esteban left the camp and went to the city of Cíbola, taking with him his army of 300 braves and many womenfolk and other followers.

It was sunset as this crowd approached the town, but the lord of Cíbola would not let them in.

The following morning, "with the sun a lance's length above the horizon," the two Indian armies faced each other in a field outside the town. But when Esteban and his principal advisers saw the Cíbolan archers taking up their positions, Esteban lost his nerve and fled as fast as he could. The archers fired and fired again. Arrows rained down on the army, and most of Esteban's men were killed.

The two Indian fugitives who told the tale explained that they had survived the massacre by hiding among the cadavers of their companions. They had not seen Esteban again, but they assumed that "he must have been shot."

Marcos's narrative offers no explanation of why he came across these two survivors only "a day's journey from Cíbola" when the massacre had happened many days before. What, one wonders, had they done in the meantime? There is no evidence in Marcos's report that Mendoza or Zumárraga forced him to answer this and other obvious questions. In fact, the official documents show no evidence that anyone in Mexico doubted the obviously inconsistent story.

Because we know that Marcos cannot have been where he said he was when heard these accounts of Esteban's defeat, and that he cannot have done so at the times he claimed he did, the whole account of the different Indian reports of a massacre is thrown into doubt. What is more, such a violent reaction was completely out of character with what we know of the Zunis. There were many important Sonorans among Esteban's entourage. Why, one is impelled to ask, would the Zunis have so readily opened hostilities with those important trading partners, who also supplied them with seasonal labor? Why attack men who came in peace?

It is surely significant that although Marcos de Niza gave the impression that Esteban was dead, he was careful not to actually assert this as fact in his report. Whatever the reality of the accounts received by Marcos of a massacre at Cíbola, and wherever he may have been when he received those reports, the friar knew that Esteban might still be alive and could therefore be found by later explorers.

MARCOS NOW BEGAN the long journey back to Culiacán and from there to Mexico City, which he reached in August or September 1539. His sensational account of his journey to the Seven Cities of Cíbola galvanized the Spanish conquistador community, and within weeks Francisco Vázquez de Coronado set out at the head of his large expedition, hoping to subjugate the northern Indians and establish colonies, missions, and cattle ranches in their wealthy lands.

At the same time, Mendoza organized a maritime expedition under the captaincy of Hernando de Alarcón to provide sea support for Coronado. So, two months after Coronado mustered his troops at Guadalajara, in February 1540, and marched for Cíbola, Alarcón sailed with two ships from the Pacific port of Santiago with orders to go along the coast of New Galicia and search for the head of the Gulf of California. Alarcón went prepared to take his expedition inland, and when he reached the mouth of the Colorado River, he launched the longboats and embarked on long exploration upstream.

Alarcón's account of his exploration up the Colorado River in 1540 and his largely friendly, albeit tentative, encounters with the Indians who lived on its banks is detailed and carefully observed. With considerable rigor, he distinguishes between his factual observations and his personal impressions, quite unlike Marcos or Cabeza de Vaca, and there is a welcome clarity to his report. He seems a reliable witness and a thorough chronicler.

Because his mission was to supply Coronado's army, Alarcón pushed the expedition to travel as far up the Colorado River as possible. As they traveled, he diligently questioned the Indians for information about Cíbola, sometimes resorting to unreliable sign language, but using interpreters whenever they were available. Alarcón reported that when they had penetrated far upstream, one of the local Indians who was acting as an interpreter noticed the expedition greyhound, an ideal dog for hunting hare and rabbit in the rough bush country. This man remarked that he had seen these strange animals before, at Cíbola. Then, as the interpreter watched the daily comings and goings of the Spaniards with a keen eye, he pointed to some colorful ceramic plates and said that he had also seen that kind of thing at Cíbola. Both the dog and the plates, he explained, had been brought by a black man who wore a beard.

For the moment, Alarcón gleaned no further information about Esteban, but as he continued up the river, he came across an Indian chief who had been at Cíbola himself. Alarcón, hoping for news of Coronado, asked this man if he had ever heard of other people like the Spaniards.

"No," he said. But then he went on to comment that "there was a black man who wore bells on his arms and legs and had a dog much like your own."

"Your Lordship," Alarcón reminded Mendoza in his report, "must remember how the black man who traveled with Fray Marcos used to wear jingle-bells and feathers on his arms and legs and carried colored plates about with him."

The Indian chief said that Esteban had been murdered at Cíbola.

"Why?" asked Alarcón.

"The Lord of Cíbola," the chief explained, "had asked the black man whether he had any brothers."

"I have an infinite number," Esteban is supposed to have replied. "They are very well armed and are close at hand."

"When the lords of Cíbola heard this news, they held a council meeting and determined that it would be best to kill the black man so that he could not report to his brothers. That is why they killed him," the Colorado Indian explained. "Then they cut him up into many pieces and these were distributed among all the lords of Cíbola so that they would know for certain that he was dead."

Years later, Castañeda reported that Coronado himself had been told that when Esteban reached Zuni, the elders and the lords of the land had questioned him about the reasons for his visit. Then, for three days they had sat in council, deliberating on how best to proceed. They considered Esteban's improbable report that he was a herald for a great white lord who had descended from heaven and were puzzled that a black man should come as a messenger for white men. They concluded that he must be a spy, and that was why they executed him.

As early as April 1540, after Coronado and Alarcón had set out, Mendoza received apparent confirmation of Esteban's death from another reconnaissance mission, but without a reliable eyewitness account or a body, the possibility remained that Esteban had survived.

Then, in August 1540, Coronado wrote to the viceroy from Zuni, noting that the expedition had found many of Esteban's possessions and explaining that he had reached the conclusion that "the death of the *negro* is perfectly certain."

RESURRECTION

Dowa Yalanne, "Corn Mountain" (2006). The Zunis took refuge on top of this sacred mountain in times of trouble. Esteban is usually thought to have met his death at the pueblo of Kiakima, which nestled at the foot of these cliffs. *(Courtesy of the author)*

Should I die in a foreign land,	*Si muero en tierras ajenas,*
Then, who will mourn for me?	*Lejos de donde nací,*
Far from where I was born,	*¿quién habrá dolor de mí?*
Where I am a stranger unknown,	*¿Quién me terná compasión*
Who will look after me?	*Donde no soy conocido?*
Oh, unhappy and captive lover,	*¡Oh triste amador perdido*
You will never again be free,	*Cautivo sin redención!*
And your homeland is lost to thee.	*Extraño de mi nación,*
Far from where I was born,	*Lejos de donde nací,*
Who will mourn for me?	*¿quién habrá dolor di mí?*

P. Girón, sixteenth-
century Spanish poet

IN LATE JANUARY 2006, I arrived in the Sonora valley in north-west Mexico aboard a battered bus filled with schoolchildren, old women, and a few chickens, and got off at the small town of Baviácora. Large white crosses stood here and there all along the side of the road that wound up through the gorge above Corazones, as if they were messages from Esteban, but in reality marking the sites of many fatal automobile accidents. The bus rattled away down a long dusty hill leading out of town. The scene was so quintessentially evocative of the Hollywood image of Mexico that had it not been for Baviácora's unquestionable authenticity, I might have felt as though I had escaped reality and become part of the fictional world.

No one near the bus stop had much idea if there was a place to stay. I crossed the plaza in the intense, dry heat and went into the relative cool inside a dark general store. The manager and owner, Miguel, sold me a bottle of beer, opened one himself, and as we slowly finished our drinks, he pointed me across the plaza, beyond the abandoned mission church, to a building opposite the town cantina, the Hotel

San Francisco. As I looked across the square, I half wondered whether I might find a French friar called Marcos in the cantina.

I spent the afternoon with Miguel, drinking a strong homemade mezcal called *bacanora* and eating delicious *elotes* made by his wife. He knew all about Esteban, and Marcos de Niza, and Cabeza de Vaca. He told me that two years before some other gringo had passed through the town, inquiring after the dead. We chatted about life in Baviácora, and in due course Miguel tried to sell me a $100,000 share in the gold mine he was going to open up on a plot of land he owned in the hills to the east. He produced a ream of papers showing geological tests and estimates of the cost of extracting the gold and talked of the rising gold price in the international market. He explained that his son worked in some relevant government department, which would greatly ease the usual inconveniences of Mexican bureaucracy.

"Why," I asked him, "are you inviting me to have a part of your business?"

"Why not?" he replied. "I need an investor and you are a gringo. A hundred thousand dollars is nothing to you."

As dusk approached, I left him to his dreams and his *bacanora* and walked up to the cemetery, on a hill above the town. A bronze light gilded the headstones, glowed on the fresh floral wreaths, and tinged the landscape with a sepia hue as dust blew up from the fields in little eddies. Up on that hill, there was a cool breeze and a magnificent view of this strange, remote valley that still promised an improbable golden future for passing Europeans willing to surrender to the magic of hope.

It was clearer to me then than it ever has been that the seed from which Marcos de Niza's account of the Seven Cities of Cíbola would grow must have been well watered while the friar rested here in this Land of Maize with its *bacanora* daydreams and promises of wealth.

I walked back down the tranquil streets filled with happy children at play and cowboy youths with fluffy mustaches and battered pickup trucks. I ate stinging hot tacos at the portable stall in the plaza and then went by a peculiar, long-abandoned church, a simple architectural fantasy in reinforced concrete, on the verge of collapse.

Tired, no longer hungry, and more or less sober I went back to my musty room in the Hotel San Francisco, spread out my books and photocopies, and turned my attention to the question of what had happened to Esteban.

IN THE SUMMER of 1540, Coronado's expedition reached the Zuni River. They were exhausted, shattered by the almost impassable road through the Mogollon Rim along which Marcos had brought them. As they marched on the six towns, they found the Zunis ready for war. The men had retreated to Kiakima, the best defended of the pueblos, and turned it into a formidable redoubt. The women and children, the old and the infirm, had all been evacuated to the natural fortress of Dowa Yalanne, the sacred sand-red mesa above the town. Only the warrior braves remained, quietly determined to defend their way of life.

The battle was hard fought at close quarters, and Coronado was knocked senseless by a cataract of rocks hurled furiously from the flat-roofed houses. Unconscious, he was dragged to safety. But the Spanish conquistadors were at their strongest in adversity and they fought with unwavering resolve until their superior weaponry breached the defenses and the men of Cíbola finally capitulated. The routed Zuni army retreated to Dowa Yalanne. Many of their bravest young men were dead.

The Zunis did not see themselves as a warlike people, and it was now clear that engaging the Spanish army had led only to ignominious defeat. The humiliated Zuni bow priests now conferred, their authority under threat. They were no doubt thankful when Coronado sued for peace and tentative embassies began to shuttle between the opposing camps. Eventually, a delegation of Zuni elders was sent to conclude the truce.

Coronado reported to Mendoza that during those negotiations, the Zunis told him that they had killed Esteban "because the Indians of Chichilticale had told them that he was a bad man." But he then added that the Zuni elders went on to explain that Esteban "was quite unlike" Coronado and his men, "who never kill women. For the truth was that Esteban went about murdering women" at whim. These assaults on the womenfolk, "whom the Indians love better than themselves," were the reason that the Zunis killed Esteban, Coronado explained.

This is evidently slander, almost certainly invented by Coronado as a way of distracting attention from the excesses of his own army. Many surviving documentary sources clearly show that after various battles with Indians, Coronado was unable or unwilling to restrain the psychopathic sexuality of many of his own men. They had raped and abused the Indian women mercilessly.

By contrast, all the documentary evidence indicates that Esteban was quite a different character from Coronado's evildoers. While it is reasonable to speculate that the four survivors' shamanism may have led them to take sexual advantage of Indian women from time to time—possibly even prolifically, and there is some evidence that Esteban happily continued that practice as he traveled north to Cíbola with Marcos—even that largely imaginary picture of Esteban as a promiscuous lover hardly fits Coronado's description of him as a sexual predator who liked to rape and butcher women as a way of life. There is a deep-seated, familiar racism about that image of Esteban as a black sexual predator, a racism which will persist as long as the image is treated as true. It is high time that this explanation of what happened at Zuni is buried once and for all.

So the Zunis' motive for Esteban's murder, as reported by Coronado, is demonstrably spurious and clearly originated with crimes committed by Coronado's men. Without that motive, the Zunis' confession, as it were, loses all credibility. In fact, Coronado betrayed his own lack of confidence in that confession when he explained to

Mendoza that he was "perfectly certain" that Esteban was dead, not because the Zunis had confessed, but because his men had "found many of the things which Esteban wore." But it is a strange kind of logic that leads directly to the conclusion that a man's possession are, in his absence, evidence of his death. It is equally likely that Esteban had simply changed his clothes, perhaps adopting the dress of his Zuni hosts.

During his peace negotiations, Coronado heard that the Zunis had living with them a lad from Petatlán, Bartolomé, who had traveled there with Esteban and who spoke the Mexican Nahuatl language. Coronado quickly recognized that this youth would make an invaluable interpreter, and he demanded that the Zunis hand him over. But the Zunis dissembled: at first they said he was dead; then they claimed that he had been taken to a nearby town, Acucu. Faced with this apparent intransigence, Coronado quickly raised the stakes and issued angry threats. The Zunis soon handed over Bartolomé.

As I looked over the evidence in the Hotel San Francisco in Baviácora, I noticed an extraordinary and possibly very significant anomaly: in an age that attached so much importance to eyewitness testimony, Coronado had failed to record Bartolomé's account of what had happened to Esteban. Coronado must have asked, but there is no record of the answer. Whatever that answer may have been, it seems almost certain that Bartolomé said nothing to confirm that Esteban was dead. In fact, it seems likely that he either said nothing at all about Esteban or said something that Coronado preferred not to record.

My cross-referencing and examination of the Spanish documents led me to the conclusion that there was insufficient evidence to bring the Zunis to trial for Esteban's death in a modern courtroom, let alone convict them of murder or manslaughter. It was time to turn to the Zunis themselves for a clue to the events of the summer of 1539.

THE NEXT DAY, at dawn, I walked back through the long shadows thrown by the early morning light across the broad plaza of peaceful

Baviácora. The sun was a lance's length above the horizon and the air was crisp with the brisk cool of daybreak which comes before the heat. I ate tacos at the stand and had coffee. Miguel's store was closed up and there were no early dreams for sale. The bus came and stopped, and I continued my slow journey up the Sonora valley.

By lunchtime, we had arrived at the small town of Arizpe, erstwhile capital of the Mexican Internal Provinces, from where, in the eighteenth century, a Spanish governor ruled all northern Mexico, California and Washington, Arizona, and New Mexico. I ate quesadillas in a roadside shack next to the narrow, homely highway and afterward sat in the shady town square to wait for the afternoon bus that would take me up onto the high and arid flatlands of Cananea. I opened my rucksack and eased the package of maps and battered photocopies from among my dirty clothes. I turned to a remarkable account of life at Zuni Pueblo during the late nineteenth century, written by a remarkable character called Frank Hamilton Cushing, "the man who became an Indian."

Cushing was an eccentric Boston aristocrat who arrived in Zuni in 1879 as a member of an anthropological expedition sent to the southwest by the Smithsonian Institution to gather information about Indian society and culture. The expedition scientists set up camp at a respectable distance from the pueblo and began to observe the comings and goings of the "natives" as they went about their daily business. But Cushing soon became dissatisfied with the haughty methods of investigation employed by his superiors, and the intransigent secrecy of the Indians. He realized that this complete lack of communication would make meaningful research impossible. So, with the courage and arrogance of his social class, Cushing ignored the express orders of his commanding officer by abandoning the expedition camp and, quite uninvited, taking up residence in the quarters of the Zuni governor. That governor, a sanguine, kindly character called Palowahtiwa discovered his unannounced and uninvited guest slinging a hammock across his rooms and generously took in the intruder, albeit with an appropriately measured lack of grace. Cushing remained at Zuni for four years, becoming part of the Zunis' world, in due course taking the requisite number of scalps to be initiated as a bow priest.

Cushing was a fine example of the frequent cultural anomalies that enriched the Victorian age. He had stoically suffered an abject childhood of alienation from his parents and peers due to his unhealthy constitution and his love of solitude. At Zuni, he now found companionship and an outsider's freedom. He had arrived begging for shelter, but he was to leave a hero who successfully championed the Zunis' cause, using a western armory of law, politics, and journalism. He fought a long and bitter struggle in Washington to correct a bureaucratic error that had annexed rich farmlands and important water sources from the Zuni reservation.

The potential parallels between Cushing and Esteban are unmistakable. Both men represented and symbolized a colonial invasion from which they themselves were spiritually divorced. Both men arrived at Zuni very much a part of those colonial forces, bringing with them a sense of danger and suspicion from which each was anxious to separate himself. Both offered themselves to the Zunis as guests, each a traveler from a far-off land, each seeking food and shelter. Each was eager to engage with the Zunis and their ways.

Caught between two cultures, but born a "good" New Englander, Cushing inevitably turned adventure into a virtue by writing about Zuni and in the process invented modern anthropological practice. Never before had a researcher subsumed himself within a native culture so as to observe and experience its every aspect, and the descriptions and images he published were unique. But by finally publicizing Zuni secrets, Cushing permanently broke his spiritual bond with the land and people who had made a valiant warrior out of the boy who first knocked hopefully at the Zuni door. He has never been forgiven for betraying the vows of silence which rule the ritual and religious world at Zuni. Cushing is remembered as a traitor.

His accounts of Zuni are consequently invaluable to the outsider. In the square at Arizpe, I now turned to my copy of his classic monograph *Zuñi Breadstuff,* in which he records the Zunis' account of how they first began to grow wheat at the pueblo.

A group of elder Zunis told Cushing that many generations before, "certain grey-robed water-daddies," Franciscan friars, "brought and planted wheat germ and taught us how to grow it." These friars had arrived at the time that "the Indians of the Land of Everlasting Summer," the Sonorans, had come to Zuni "with long bows and cane arrows." They were accompanied by "Black Mexicans," who carried "thundering sticks which spit fire," and they were dressed in "coats of iron." But "our ancient ancestors at Kiakima" then "greased their war clubs with the brains of the first of those Black Mexicans."

What "bad-tempered fools!" the Zuni elders lamented of their ancient ancestors. They explained to Cushing that as a consequence, the "mustachioed" Spaniards grew angry and "appeared in fear-making bands and grasped the life-trails of our forefathers until they became like dogs after a drubbing."

Cushing came to the conclusion that this first "Black Mexican" must have been Esteban, as he explained in a lecture he gave to the Geographic Society of Boston. He arrived to give the talk dressed in his own interpretation of Zuni costume, just as Esteban wore his shaman's accoutrements when he was in Mexico. Cushing began with the theatrical opening of a natural dramatist.

"Like the teller of Indian tales," he began, "I bid you back more than three hundred and fifty years."

An able raconteur, Cushing now carried his audience from the elegant eastern hall of learning to the firelight world of a distant evening in his little room at the heart of Zuni Pueblo. His audience soon felt as though they were there at his side as he sat reading an "old work of travel." They could almost see the four tribal elders who entered the semi-European enclave of his room and sat down to roll cigarettes. As they smoked, they watched, until one man spoke up, talking as if for the rest.

"What do the marks on the paper-fold say?" the old men wanted to know.

"Old things," was all that Cushing was prepared to reply.

"How old?" they asked.

"Three hundred and fifty years."

"How long is that?"

The four old men could not conceptualize such a long passage of time. But they were intellectually resourceful and slowly laid out a chain of 350 corn kernels, "in a straight line across the floor."

They huddled over, counting the kernels and remarking here and there, "Now that's one father. This is his son," marking out the generations.

Eventually they established that ten or eleven generations were roughly equivalent to Cushing's 350 years.

"Why!" they exclaimed. "That must have been when our ancients killed the Black Mexican at Kiakima."

"Tell me more," said Cushing.

"It is to be believed that a long time ago, when roofs lay over the walls of Kiakima, when smoke hung over the house-tops, and the ladder-rounds were still unbroken—It was then that the Black Mexicans came from their abodes in Everlasting Summerland. One day, unexpected, out of Hemlock Canyon, they came, and descended to Kiakima. But when they said they would enter the covered way, it seems that our ancients looked not gently on them. But with these Black Mexicans came many Indians of Sonoli, as they call it now, who carried war feathers and long bows and cane arrows like the Apaches, who were enemies of our ancients. Therefore these our ancients, being always bad tempered and quick to anger, made fools of themselves after their fashion, rushed into their town and out of their town, shouting, skipping and shooting with sling-stones and arrows and war clubs. Then the Indians of Sonoli set up a great howl, and they and our ancients did much ill to one another. Then and thus, the black Mexican, a large man with Chilli lips, was killed by our ancients right where the stone stands down by the arroyo of Kiakima. Then the rest ran away, chased by our grandfathers, and went back toward their country in the Land of Everlasting Summer. But after they had steadied themselves and stopped talking, our ancients felt sorry, for they thought, 'Now we have made bad business, for after a while, these people, being angered, will come again.' So they felt always in danger and went about watching the bushes. By and by they did come back, those Black Mexicans, and with them many men of Sonoli. They wore coats of iron and even bonnets of metal and carried for weapons short canes that spit fire and made thunder. Thus it was in the days of Kiakima."

Cushing told his Boston audience that he was still utterly ignorant of the Spanish histories of Marcos and Esteban when he heard this story from the Zunis. But we know this was not strictly true, because he had written about Marcos, and Coronado, and Cabeza de Vaca, and other ancient chroniclers of the Spanish Empire in letters he had sent to Washington long before he took up residence in Zuni Pueblo. I wondered how much Cushing had elaborated his account and to what extent he had prompted the old Zuni historians into telling the story he expected to hear rather than the story that they usually told.

I also wondered how much Zuni oral tradition had been influenced by alien accounts of their history. For much of modern history, Zuni has been constantly visited by outsiders: anthropologists, ethnographers, archaeologists, and Anglo academics looking for their own long-lost souls; by Spanish Catholics looking for the lost souls of the

Zunis, and by Protestant priests looking to do good. What imprint did these foreigners leave behind them with their constant questions, their constant quest for the intangible truth, and their constant quest to educate?

By way of explanation, it is instructive to tell a story that took place high up in the Andes of Peru, where the Incas built their awesome palaces and temples of vast and irregularly shaped stone building blocks that were so smoothly and perfectly cut that it is impossible to fit even a razor blade between them. Over the years, thousands of archaeologists had tried, without success, to work out how the Incas managed to do this. Then, one archaeologist discovered at an important Inca site the remains of some kind of saw for cutting stone. He quickly worked out how the contraption was meant to work and built one for himself, convinced that he had solved this ancient problem. For a time, his colleagues believed that he had, until they all realized that he had in fact found the remains of an experiment discarded by a nineteenth-century scholar who had also hoped to solve the mystery.

Had the Zunis been so often asked about each documentary detail of Esteban's death by western scholars that those details had become part of the Zunis' landscape, just as the discarded saw fleetingly became a part of Inca achitectural history?

It is interesting that both Zuni accounts of the "Black Mexicans" that were recorded by Cushing emphasize a sense of the ancestors' stupidity in killing the first Black Mexican. The nineteenth-century Zuni elders, it seems, saw this aggressive behavior as out of character, for, as all the books had led me to expect, the Zunis saw themselves as traders who relied on peaceful relations with their neighbors and preferred to avoid conflict. Understood as a parable rather than history, the Zuni story of the Black Mexican clearly preaches a message of peace, for their unprovoked aggression is shown to lead directly to the arrival of the angry Spaniards. And the truth of a parable, of course, is not in the narrative itself, but in the morality of the tale.

I was to discover that even today, Zunis tend to show incomprehension and a sense of disbelief when faced with Spanish accounts of how their ancestors murdered Esteban. It is not a story that Zunis are comfortable with, almost as though it does not belong to their own sense of history.

Among my travel-worn photocopies, I had an account of Esteban's death which was published in 1997 and which had been given by a Zuni scholar and historian called Edmund Ladd as a spontaneous contribution to an academic conference on the Coronado expedition. *Zuni on the Day the Men in Metal Arrived* is a masterpiece of oral academia of a kind all but lost in the e-mail world of modern universities.

Ladd told the story of Esteban's arrival at Zuni, but while he gave the impression that he was working within an oral tradition, he also pointed out that oral traditions lack detail and that his account was in fact based on "other" sources, tempered by Zuni spiritual and moral values. In reality, Ladd, it seems, needed to rely on Spanish documents for the detail. And so, with evident reluctance, he found himself forced to repeat the explanation for Esteban's death that had been offered by Castañeda in 1540. He concluded that Esteban's fatal mistake was to declare himself a leader of the Spaniards and that he must therefore have been killed because he was believed to be a spy working for an evidently powerful enemy. But Ladd was clearly puzzled that the Zunis should have committed the murder at all, and he insisted that they would have been friendly at first, "for it is their nature to be" so. If the Zunis had been guilty of the killing, their descendants seemed less than convinced.

I looked up from my books. Children were fooling around in the street outside the closed-up Cathedral of Arizpe and it was very hot and the town was asleep. It reminded me of Spain at siesta time in the summer. I chatted with a shoeshine boy while he shined my shoes. He refused payment, but I insisted and so he sat down once more and polished them all over again.

At exactly two o'clock, precisely on time by the town hall clock, the afternoon bus arrived and I left Arizpe. We drove up and out of the Sonora valley, leaving the ancient Zunis' "Land of Eternal Summer" for the flat deserts of the border country. There were cactus and purple prickly pears, and a warm breeze blew in through the open door at the front. That night I slept in the unpleasant border city of Nogales in a cheap hotel and drank tequila in a disagreeable bar full of the drunken dregs of some fraternity house. It was my last night in Mexico.

THE FOLLOWING MORNING, I passed the lines of nervous pedestrians with their cheap luggage full of cheap medicines, walked across the border into the United States of America, and found a van full of Mexicans who would take me to Tucson. The driver was a graduate of Guadalajara University with a particular interest in British constitutional history. As he drove us through the back streets of Tucson's Mexican barrios and industrial estates, dropping off the other passengers, he talked to me about Magna Carta, the Glorious Revolution, the founding of the United States, universal suffrage, and the principles of the Common Law. When we reached my destination, the Arizona University law school, I stayed chatting for a while before we shook hands and I shouldered my filthy rucksack. Dirty, tired, and happy, I went into the air-conditioned modern building and looked for Jim Anaya, the Indian lawyer who specialized in the rights of indigenous peoples whom I had met in Seville. He had promised to take me to Zuni.

As I sat in the law school office, watching Jim and his friend Rob Williams faxing and phoning their various contacts on my behalf, I noticed a slight sense of frustration. I felt trapped as I watched the impatient cultural machinery of the modern world clash with the slow, measured civility of Zuni society. But it was a fleeting sense of unease that evaporated in the sunshine of the following morning as the three of us set out on the long drive to Zuni.

As we traveled, I briefed Jim and Rob about Esteban. I told them about Marcos de Niza, the unreliable witness; and about the various reports of Esteban's death in the other Spanish accounts. I read key passages from some of the accounts and summarized others. Both Jim and Rob sit as tribal judges within the Indian legal system. If anyone could pass judgment on the charge of Esteban's murder which History had leveled at the Zunis, they could.

We drove up the Zuni valley and reached the modern pueblo, a pretty, scruffy place, full of life, and pulled into the parking lot in front of the tribal council offices. Inside, by extraordinary good fortune, we met a member of the tribal council called Edward Wemytewa, who has worked hard to keep oral tradition alive among the young people of the tribe, increasingly influenced by television and drawn to the more material world beyond their reservation. He also knew Zuni history, which made him wary, determined to engage with the outside world on his own terms. He seemed torn between silence and enthusiasm. But he also knew a lot about Esteban, and any historian is always pleased to meet a colleague. So, with reservations, he seemed happy to talk. It helped that he seemed to have mistaken Jim for somebody else.

We sat around a low coffee table in the entrance hall of the tribal council offices and Ed slowly began to tell us what he knew about Esteban's arrival at Zuni. He engaged easily with Rob and Jim, while managing to welcome and exclude me at the same time. I was glad that I had briefed Jim and Rob in the car, because I was out of my depth. As I watched, their legal training and familiarity with Indian law quickly showed in a subtle cross-examination.

Ed is a master storyteller with a keen academic mind, and his tale of Esteban's death at Zuni was filled with detail and told with rare clarity. But as Ed spoke, I silently noted that almost every fact in his story could be explained with reference to one or another of the Spanish chronicles. Nothing he said was new to me, except for one curious and emotive symbol. In passing, Ed associated Esteban with the image of a snake. This was the one detail I had never come across in my research, the only Zuni secret that Ed let slip.

When Ed had finished speaking, I tried to press him about his reference to a snake and asked him whether there was some secret about Esteban that he could not tell me, something kept hidden in the Zunis' histories and traditions. But instead he told me about the night dances the Zunis celebrate in winter, when the different kiva societies perform rituals during which the men dress as ancestral characters from Zuni history and the spirit world. As they dance, the spirits come to these actors and the supernatural world becomes fleetingly tangible.

These spirit-dancers are known as kachinas, and Ed told me about a "beautiful black kachina" with a sexual magnetism that cast a spell over the women who saw the dance. He implied that he thought the Zunis might have killed Esteban out of straightforward sexual jealousy and not because he mistreated their women. I had already learned about the black kachina Chakwaina, the monster kachina who is common to all the different pueblo cultures. Anthropologists report that a number of pueblo legends associate Chakwaina with Esteban and some have implied that in this way Esteban's spirit lives on in the pueblo world. It was beginning to seem that the Zunis may well have killed Esteban, but that he had then been resurrected as a potent kachina. Yet, even as Ed began to suggest that the black kachina might offer an explanation of what had happened to Esteban, he immediately withdrew the idea and refused to confirm that the black kachina was really Esteban's spirit at all. I was skeptical.

Later, we met Jim Anaya's old Zuni friend Jim Enote, who showed us around the house he was building on the outskirts of the modern pueblo. We sat chatting in a big white room, which felt cool and refreshing after the stuffy council building and the relentless heat outdoors. I explained to Jim that I was trying to find out about the Zunis' history of what had happened to Esteban and I told him that Ed's account seemed to be based on the Spanish sources. Then I asked him about the black kachina.

He began to tell us that the black kachina had nothing to do with Esteban, but then stopped and asked, slightly surprised, with a twinkle in his eye, "Did Ed tell you that?"

Jim Enote then talked diplomatically about Ed's expertise as a historian and his knowledge of Esteban's story, but quickly turned the conversation to a conference he had recently attended in Canada and the problems he had with funding his work on making Indian cultural maps of North America. I started to believe that Jim Enote had not only given away his own view that the black kachina had little or nothing to do with Esteban but had also let slip his surprise that Ed should have done so.

As I reflected on our conversation, I thought I understood what Ed had really been doing when he told me about the black kachina. I suspected that as he knew that I had come to Zuni for a story, and as a storyteller himself, he was glad to offer me a story to tell. He wanted to give me what I had come for. But as a historian and a protector of Zuni traditions, he gave me a story I could have found elsewhere and then refused to certify it as true. Ed kindly gave me what I needed, closure for Esteban and an end for a book, and then left it to me to decide whether the story was true. It would have been so easy to poetically resurrect Esteban as the black kachina Chakwaina here, now, on these pages, and then leave it at that.

Then, as I thought more about Ed's account of Esteban's death, his murder at Zuni hands, I became convinced that like Ladd and the Zuni storytellers described by Cushing, Ed too had been struggling to find a motive for an act that was completely out of character for his culture.

The Zunis had given a deeply symbolic meaning to their story of Esteban's murder by directly associating it with their terrible defeat by Coronado. It had become a classic example of the way in which victims of savage oppression so often try to tell their story so as to blame themselves, because through that guilt they can regain a sense of control over their own destiny. So long as the Zunis believe that their gods had punished their forebears for Esteban's murder by sending Coronado and other Spaniards to conquer them, they can still sense the part they played in their own downfall. But if Coronado's coming was not their responsibility, but was simply fate, then their only conclusion must be that their gods had abandoned them. And how can man die more miserably than amid the ravaged houses of his ancestors and forsaken by his gods?

And so, even as Edward Wemytewa told us his story, that very process of telling it, it seemed to me, was leading him to trust the Spanish sources less and less. Caught between two American cultures, Zuni and Anglo, Ed is an erudite man for whom the western historical tradition is as important a source of knowledge about Zuni history as Indian oral history should be to western scholars interested in America's past. By the tenets of either tradition and their different ways of understanding the past, Esteban's arrival marks, with unusually vivid symbolism, a critical point in Zuni history in particular and an important moment in American history in general. Of such historical moments are myths and legends born. So, like Ladd before him, Ed too had turned to Spanish sources to find out more about this ancient event, and he too seemed uncertain of what to make of those sources.

Ed, I know, guards some secret about Esteban safely preserved within the tight-lipped community of a kiva society or the Zunis' priesthoods. But those privy to such things are sworn to secrecy, and so such knowledge has no place in print. From what little I have learned about the complex world of the Zunis, it seems to me that the only plausible reason they might have had for executing Esteban was that they considered him a sorcerer. Under traditional Zuni law, only sorcery was considered a capital crime.

That afternoon Jim, Rob, and I drove out to Dowa Yalanne. The great sand-pink mesa rises steeply from the grassy river valley, dominating the landscape with its crags and buttresses. The remains of Kiakima are clustered at the foot of a wide, sweeping amphitheater, where two towers of sheer rock stand like sentinels over sacred Zuni shrines, indistinctly marked by prayer sticks among the ruins of the former pueblo. The walls of the "covered way" that led through Kiakima, described by Cushing's informants, can still be seen, a few vertical slabs that stand like tombstones marking the place where the legends claim Esteban died. It is a place of extraordinary spiritual calm.

In the evening, after this peculiar pilgrimage to the possible site of Esteban's unmarked grave, the three of us sat in a shack at the side of the main road eating tacos and burritos and drinking colas. We talked about Esteban, Ed, and the haunting tranquillity of Kiakima, and we began to wonder what to do before bedtime. Then the waitress told us that the whole town was at a basketball game and so after supper we drove to the high school. The parking area overflowed into the surrounding desert and Jim heaved the car up onto a verge near a fence. After some searching about, we found a back entrance to the gymnasium and walked into the middle of the basketball game. There were still a few empty seats halfway up one of the stands, and we picked our way through the steep terraced rows, packed with screaming supporters.

I know nothing about basketball, but Jim had played in college and pointed out the unorthodox tactics necessarily used by the Zuni team because their players were all short compared with their Navajo opponents. They rushed the length of the court with cannonball vigor, relying on a low center of gravity to outwit their gangling opponents. Sadly, despite their raw enthusiasm and evident skill, by the third quarter it was obvious that the size and power of the Navajos were prevailing and the home crowd succumbed to a growing sense of disappointment.

That disappointment was soon vented in angry outbursts directed at the referee. The atmosphere gradually intensified and as it did, Rob was riveted. "I guess this is what happened to Esteban," he said. "All those anthropologists talk about how the Zunis are peaceful and not emotionally expressive. But these guys are screaming for blood!"

Cultural comparisons are easily misleading, and I had little idea how to properly qualify or quantify the aggression shown by the Zuni spectators at this basketball game. But compared with the murderous tribal hatred of European and South American soccer fans, which often leads to physical violence, the crowd at Zuni seemed passionately benign. I was still not convinced that the Zuni bow priests had murdered Esteban.

In fact, the cultural importance attached by the Zunis to hospitality and the avoidance of conflict raised an intriguing question. Why, I wondered, when Coronado first approached them with his thirsty, hungry, fatigued, weakened army, did the Zunis—again uncharacteristically—resort to conflict rather than dialogue?

The arrival of many strangers is always good cause for apprehension, but it is an unlikely reason for a thoughtful, considered, self-confident nation with a preference for peace to suddenly go to war without provocation. For the Zunis to attack Coronado, they must have had very good reason to think it was the right thing to do.

Very very few men familiar with the Spaniards' true nature can have visited the six Zuni villages before Coronado. Perhaps only one, Esteban. Perhaps, it occurred to me, Esteban warned his hosts at Zuni what to expect from the Spaniards. Did he then move on farther north, away from the relentless advance of the Spanish frontier? Or did he stay at Zuni and urge the "bow priests" to engage Coronado in battle? Did he believe that they had a chance of victory and tell them to strike while the Spaniards were at their most vulnerable?

It would be easy to believe that in the aftermath of defeat, the bow priests turned on the African outsider. Perhaps they chased him from their midst, blaming him for their downfall. Perhaps they charged him with sorcery and had him executed?

As I WRITE these inconclusive final words, I must be flying more or less over Zuni. It is a strange and disappointing feeling. I have told you how Alvar Núñez Cabeza de Vaca ended up in Spain, disgraced after he failed as governor on the River Plate. We know that Andrés Dorantes de Carranza ended his days as a farmer and innkeeper on the road from Veracruz to Mexico City. We even know that Alonso del Castillo Maldonado was a landowner who lived in Puebla de los Angeles and Marcos de Niza retired to an alcoholic dotage in Xochimilico. But I find it more and more difficult to believe that Esteban was killed at Zuni and I have no idea what really happened to him. Just as his origins in Africa are obscure, so too his death in America has proved to be a mystery.

My flight took off from Phoenix, symbolically enough; and it may be that Esteban can rise from his Zuni ashes and reach towards the heavens, which he had so often claimed were his home. I have the transcript of a fragment from a lost and unpublished chronicle buried in the Mexican archives. It records that "when Esteban reached the Mayo River, he was so taken with the handsome and beautiful women of the Mayo Indians that he hid himself away" from Marcos de Niza and the Spaniards. There, "following the custom of the country, he took four or five wives." By 1622, "one of Esteban's sons had become a captain or lord of certain lands on the banks of the Mayo River, in the municipal district of Tesio." He was of "noticeably mixed race, tall and lean, with sallow features," a man called "Aboray."

BIBLIOGRAPHY OF SOURCES

ESTEBAN'S BIOGRAPHY

THERE ARE THREE main sources for Esteban's biography, all written by Europeans who were his companions during his adventures.

The General and Natural History of the Indies, by Gonzalo Fernández de Oviedo y Valdés, Spanish Historian Royal, which contains an account of Pánfilo Narváez's expedition to conquer and settle Florida of 1527 to 1536.

Shipwrecks, an account of that expedition attributed to Alvar Núñez Cabeza de Vaca, who survived alongside Esteban.

Marcos de Niza's *Report* on the expedition to the Seven Cities of Cíbola, modern Zuni in New Mexico, during which Esteban disappeared.

DOCUMENTARY SOURCES,
ARCHIVO GENERAL DE INDIAS (AGI)

Most of this material can now be viewed online at www.pares.mcu.es.

Indiferente 419 L 4., 106r–107v, 138v (career of Jerónimo de Alanís).

Indiferente 420 L 8, 193r–193v (as above).

Indiferente 421 L 11, f.77v (Vasco Porcallo, *regidor*).

Indiferente 421 L 12, 50v–50r 1527 (permission for black slaves).

Indiferente 421 L 12, 71v–72r 1527 (as above).

Indiferente 421 L 13, 105r–106r (as above).

Indiferente 421 L 13, 60r, 1527/1528 (Juan Suárez as Bishop of Florida).

Indiferente 423 L 20 (Juan Garrido).

Indiferente 427 R 30 (Africans as landowners).

Indiferente 1962 L 5, ff.273v–267v (Cabeza de Vaca's return to Spain).

Justicia 1173 N 5 (Juan Garrido).

Lima 565 L 3, 161v (Dorantes's stepdaughter's inheritance).

México 95 N 39 (Lope de Samaniego to crown).

México 204 N 3 (Juan Garrido).

México 212 N 45 (Baltasar Dorantes de Carranza; working transcription: www.ems.kcl.ac.uk/content/proj/disc/cab/agi/index.html).

México 1088 L 3, 251r–251v (Mari Hernández).

Patronato 20 N 5 R 10 (Marcos's account).

Patronato 55 N 5 R 4 (Juan Durán).

Patronato 57 N 4 R 1 (Alonso del Castillo Maldonado).

Patronato 65 N 1 R 4 (Antonio de Aguayo).

Patronato 157 N 2 R 4 (Francisco Díaz).

Patronato 184 R 27 (Mendoza to Crown, 1537).

Patronato 275 R 39 (Castillo marries widow).

Patronato 278 N 2 R 230 (Castillo as settler and landowner).

BIBLIOTECA COLOMBINA

Oviedo y Valdés, Gonzalo Fernández de, manuscript of *Historia*, shelfmark 57–5–43.

PUBLISHED SOURCES

REFERENCES IN THE Notes (beginning on page 393) are based on the author-date system. However, it seems absurd in this context to give the date of modern editions of works written many years ago. Therefore, references in the Notes sometimes show a date in square brackets (e.g. "Las Casas [1552]" or "Navagero [c. 1526]"), referring to either the date of first publication if it was during the author's lifetime or soon afterward, or the approximate date that the work was written. The actual editions used are indicated in the list below:

Adorno, Rolena. "The Discursive Encounter of Spain and America: The Authority of Eyewitness Testimony in the Writing of History," *William and Mary Quarterly* 49:2 (1992), 210–228.

Adorno, Rolena, and Charles Patrick Pautz. *Alvar Núñez Cabeza de Vaca: His Account, His Life, and the Expedition of Pánfilo de Narváez,* 3 vols. (Lincoln and London: University of Nebraska Press, 1999).

Aguirre Beltrán, Gonzalo. *La población negra de México: Estudio etno-histórico,* 2nd ed. (México: Fondo de Cultura Económica, 1972 [1946]).

Aguirre Beltrán, Gonzalo. *Medicina y magia: El proceso de aculturación en la estructura colonial. Obra antropológica VIII* (Mexico DF: Universidad Veracruzana, 1992 [1963]).

Aiton, Arthur S., and Agapito Rey. "Coronado's Testimony in the Viceroy Mendoza *Residencia,*" *New Mexico Historical Review* 12:2 (1937), 288–329.

Algería, Ricardo A. *Juan Garrido: El conquistador negro en las Antillas, Florida, México y California, c. 1503–1540* (San Juan de Puerto Rico: Centro de Estudios Avanzados de Puerto Rico y el Caribe, 1990).

Armas Wilson, Diana de. "Refashioning a Gentleman: Cabeza de Vaca in Captivity," in *Voces a Ti Debidas: In honor of Ruth Anthony el Saffar,* ed. Maria Cort Daniels et al. (Colorado Springs: Colorado College, 1993), 21–28.

Arens, William. *The Man-Eating Myth* (Oxford: Oxford University Press, 1978).

Arrom, José. "Gonzalo Fernández de Oviedo, Relator de Episodios y Narrador de Naufragios," *Ideologies and Literature* 17:4 (1983), 133–145.

Arteaga, Armando. "Fray Marcos de Niza y el Descubrimiento de Nuevo México," *Hispanic American Historical Review* 12:4 (1932), 481–492.

Benítez, Fernando. *Los primeros Mexicanos: La vida criolla en el siglo XVI* (Mexico DF: Ediciones ERA, 1962 [1953]).

Bennett, Herman L. *Africans in Colonial Mexico: Absolutism, Christianity, and Afro-Creole Consciousness, 1570–1640* (Bloomington and Indianapolis: Indiana University Press, 2003).

Bloom, Lansing. "Who Discovered New Mexico?" *New Mexico Historical Review* 15 (1940), 101–132.

Bloom, Lansing. "Was Fray Marcos a Liar?" *New Mexico Historical Review* 16 (1941), 244–246.

Blumenthal, Debra. "'La Casa dels Negres': Black African Solidarity in Late Medieval Valencia," in *Black Africans in Renaissance Europe,* ed. T. F. Earle and K. J. P. Lowe (Cambridge: Cambridge University Press, 2005), 225–246.

Bolaños, Alvaro Félix. "Panegírico y libelo del primer cronista de Indias, Gonzalo Fernández de Oviedo," *Thesaurus: Boletín del Instituto Caro y Cuervo* 45:3 (1990), 577–649.

Boyd-Bowman, Peter. *Indice geobiográfico de más de 56 mil pobladores de la América hispánica,* 5 vols. (México: Fondo de Cultura Económica, 1985).

Boyd-Bowman, Peter. *Indicé y extractos del Archivo de Protocolos de*

Puebla de los Ángeles, México (1538–1556) (Madison: Microfilm, 1988).

Cabeza de Vaca, Alvar Núñez. *Naufragios*, ed. Juan Maura (Madrid: Cátedra, 2001 [1542]).

Cadamosto, Avise da. *The Voyages of Cadamosto; and Other Documents on Western Africa in the Second Half of the Fifteenth Century*, trans. and ed. G. R. Crone (London: Hakluyt Society, 1937 [1507]).

Callender, Charles, and Lee M. Kochems. "The North American Berdache," *Current Anthropoogy* 24:4 (1983), 443–470.

Castellano, Juan R. "El negro esclavo en el entremés del Siglo de Oro," *Hispania* 44 (1961), 55–65.

CDI Colección de documentos inéditos relativos al descubrimiento, conquista y colonización en América y Occeanía, ed. Joaquin F. Pacheco and Francisco de Cárdenas, 42 vols. (Madrid: Real Archivo de Indias, Bernaldo de Quirós, 1864–1884).

Cervantes, Fernando. "The Idea of the Devil and the Problem of the Indian: The case of Mexico in the Sixteenth Century," Institute of Latin American Studies, Research Papers (London: University of London/Institute of Latin American Studies, 1991).

Cervantes de Salazar, Francisco. *Life in the Imperial and Loyal City of Mexico in New Spain*, ed. Carlos Eduardo Castañeda, trans. Minnie Lee Barrett Shepard (Austin: University of Texas Press, 1953 [1554]).

Chaves, Manuel. *Cosas nuevas y viejas: Apuntes sevillanos* (Seville: Sauceda 11, 1904).

Chimalpahin Cuauhtlehuanitzin, Domingo de San Antón Muñon. "Diario," in *Lecturas Mexicanas*, ed. Ernesto de la Torre Villar, 5 vols. (Mexico DF: Universidad Nacional Autónoma de México, 1998).

Cortés López, José Luis. *La esclavitud negra en la España peninsular*

del siglo XVI. Acta Salmanticensia: Estudios históricos y geográficos 60
(Salamanca: Universidad de Salamanca, 1989).

Cortés, Vicenta. *La esclavitud en Valencia durante el reinado de los
Reyes Católicos (1479–1516)* (Valencia: Ayuntamiento, 1964).

Cortés, Vicenta. "Valencia y el comercio de esclavos negros en el
siglo XV," *Studia (Lisbon)* 47 (1989), 81–145.

Covarrubias Orozco, Sebastáin de. *Tesoro de la lengua castellana o
española*, ed. Felipe C. R. Maldonado and Manuel Camarero
(Madrid: Castalia, 1995 [1611]).

Cuevas, Mariano P., and García Genero, eds. *Documentos inéditos
del siglo XVI para la historia de México* (Mexico: Porrúa, 1975).

Cushing, Frank H. "The Discovery of Zuni, or the Ancient Provinces of Cibola and the Seven Lost Cities" (lecture given to the
Geographic Society of Boston in 1885), in *Zuñi: Selected Writings*,
ed. Jesse Green (Lincoln and London: University of Nebraska
Press, 1979), 171–175.

Cushing, Frank H. *Zuñi Breadstuff: Indian Notes and Monographs*,
vol. 8 (New York: Museum of the American Indian and the Heye
Foundation, 1920).

Daza, Juan. *Estracto de las ocurrencias de la peste que aflixió a esta ciudad
(Jerez de la Frontera) en el año 1518 hasta el de 1523, por Juan Daza*,
ed. Hipólito Sancho (Jerez de la Frontera: Publicaciones de la
Sociedad de Estudios Históricos Jerezanos, Primera Serie, 1, 1938
[c. 1523]).

Díaz del Castillo, Bernal. *Historia verdadera de la conquistad de la
Nueva España*, 3 vols. (Mexico DF: Pedro Robredo, 1944 [c. 1568]).

Dobyns, Henry. "Estimating Aboriginal American Population: An
Appraisal of Techniques with a New Hemispheric Estimate,"
Current Anthropology 7 (1966), 395–416.

Dobyns, Henry. *Their Number Become Thinned: Native American
Population Dynamics in Eastern North America* (Knoxville: University of Tennessee Press, 1983).

Domínguez Ortiz, Antonio. *La esclavitud en Castilla en la edad moderna y otros estudios de marginados* (Granada: Comares, 2003).

Dorantes de Carraza, Baltasar. *Summaria relación de las casas de la Nueva España* (Mexico D.F.: Porrúa, 1987 [1604]).

Driver, Harold E. *Indians of North America*, 2nd ed. (Chicago: University of Chicago Press, 1969 [1961]).

Elvas, Fidalgo de. *Expedición de Hernando de Soto en la Florida*, trans. Miguel Muñoz de San Pedro, 3rd ed. (Madrid: Espasa-Calpe, 1965 [c. 1540]).

Epstein, Jeremiah F. "Cabeza de Vaca and the Sixteenth-Century Copper Trade in Northern Mexico," *American Antiquity* 56:3 (1991), 474–482.

Fernandes, Valentim. *Códice Valentim Fernandes*, ed. José Pereira da Costa (Lisbon: Academia Portuguesa da História, 1997 [c. 1507]).

Flint, Richard. *Great Cruelties Have Been Reported: The 1544 Investigation of the Coronado Expedition* (Dallas: Southern Methodist University Press, 2002).

Flint, Richard, and Shirley Cushing **Flint**, eds. *Documents of the Coronado Expedition, 1539–1542: "They Were Not Familiar with His Majesty, nor Did They Wish to Be His Subjects"* (Dallas: Southern Methodist University Press, 2005).

Fonseca, Jorge. "Black Africans in Portugal during Cleynaert's Visit (1533–1538)," in *Black Africans in Renaissance Europe*, ed. T. F. Earle and K. J. P. Lowe (Cambridge: Cambridge University Press, 2005), 113–121.

Franco Silva, Alfonso. "Los negros libertos en las sociedades andaluzas entre los siglos XV al XVI," in *De l'esclavitud a la llibertat: Escaus i lliberts a l'edat mitjana. Actes del Co.loqui Internacional celebrat a Barcelona, del 27 al 29 de maig de* 1999 (Barcelona: Consell Superior d'Investigacions Científiques, 2000), 573–592.

García Icazbalceta, Joaquín. *Don Fray Juan de Zumarrága: primer* *obispo y arzobispo de México*, 4 vols., ed. Rafael Aguayo Spencer and Antonio Castro Leal (Mexico: Porrúa, 1947).

García Icazbalceta, Joaquín, ed. *Nueva colección de documentos para la historia de México*, 5 vols. Vol. 1: *Cartas de religiosos de Nueva España: 1539–1594* (Mexico: Antigua Librería de Andrade y Morales, 1886).

García Icazbalceta, Joaquín. "Introduction," in Fernán González de Eslava, *Coloquios Espirituales y Sacramentales y Poesías Sagradas* (Mexico: Antigua Librería, 1877 [1610]).

Gatschet, Albert S. *The Karankawa Indians: The Coastal People of Texas*, with notes by Alice W. Oliver and Charles A. Hammond; Archaeological and Ethnological Papers of the Peabody Museum, 1:2 (Cambridge, MA: Peabody Museum of American Archaeology and Ethnology, 1891).

Gerbi, Antonello. *La naturaleza de las Indias nuevas: De Cristóbal Colón a Gonzalo Fernández de Oviedo* (Mexico: Fondo de Cultura Económica, 1978).

Gestoso y Pérez, José. *Curiosidades antiguas sevillanas*, Segunda Serie (Seville: Correo de Andalucía, 1910).

Gibson, Charles. *The Aztecs under Spanish Rule: A History of the Indians of the Valley of Mexico, 1519–1810* (Stanford, CA: Stanford University Press, 1964).

Gilman, Stephen. "Bernal Díaz del Castillo and *Amadís de Gaula*," *Studia Philologica: Homenaje ofrecido a Dámaso Alonso por sus amigos y discípulos con ocasión de su 60° aniversario, II* (Madrid: Gredos, 1961), 99–114.

Girón, P. *Crónica del emperador Carlos V*, Carta de Alonso Enríquez a Cristóbal Mejía from Santa Olalla, a 20 de marzo de 1530 (Madrid: Aguilar, 1964 [1530]), 160–161.

Gómara, Francisco López de. *Historia de las Indias y conquista de Nueva España* (Madrid: Biblioteca de Austores Españoles, 1946 [1552]).

González Rodríguez, Luis, and María del Carmen Anzures y Bolaños, "Martín Pérez y la etnografía de Sinaloa a fines del siglo XVI y principios del siglo XVII," *Estudios de Historia Novohispana* 16 (1996), 171–213.

González, Javier Roberto. "*Mal Hado—Malfado*: Reminiscencias del *Palmerín de Olivia* en los *Naufragios* de Álvar Núñez Cabeza de Vaca," *Káñina* 23:2 (1999), 55–66.

Goodwin, R. T. C. " 'De lo que sucedió a los demás que entraron en las Indias': Álvar Núñez Cabeza de Vaca and the Other Survivors of Pánfilo Narváez's Expedition," *Bulletin of Spanish Studies* 84:2 (2007), 147–173.

Goodwin, R. T. C. "Food, Art, and Literature in Early Modern Spain: The Representation of Food in Velázquez's *Bodegones, Guzmán de Alfarache, Don Quijote,* and the Still-Life Paintings of Sánchez Cotán," PhD thesis (University of London, 2001).

Graullera Sanz, Vicente, *La esclavitud en Valencia en los siglos XVI y XVII* (Valencia: Consejo Superior de Investigaciones Científicas, 1978).

Greenleaf, Richard E. "The Mexican Inquistion and the Indians: Sources for the Ethnohistorian," *Americas* 34 (1978), 315–344.

Greenleaf, Richard E. *Zumárraga and the Mexican Inquisition, 1536–1543* (Washington DC: Academy of American Franciscan History, 1961).

Gutiérrez de Medina, Cristóbal. *Viaje del Virrey Marqués de Villena* (Mexico: Imprenta Universitaria, 1947).

Hallenbeck, Cleve. *Alvar Núñez Cabeza de Vaca: The Journey and Route of the First European to Cross the Continent of North America, 1534–1536* (Glendale: Clark, 1940).

Hallenbeck, Cleve. *The Journey of Fray Marcos de Niza* (Dallas: University Press, 1949).

Hallenbeck, Cleve. *Land of the Conquistadors* (Caldwell: Caxton, 1950).

Hammond, George P., and Agapito **Rey,** eds. *Narratives of the Coronado Expedition,* 1540–1542, Corondado Cuarto-Centennial Publication 1540–1542, 12 vols. (Albuquerque: University of New Mexico Press, 1940).

Hammond, George P., and Agapito **Rey,** eds. *The Rediscovery of New Mexico,* 1580–1594. Coronado Cuarto Centennial Publications, 1540–1940 (University of California Press, 1966).

Hartmann, William K. "Pathfinder for Coronado: Reevaluating the Mysterious Journey of Marcos de Niza," in *The Coronado Expedition: The 1540–1542 Route across the Southwest,* ed. Richard Flint and Shirley Cushing Flint (Niwot: University Press of Colorado, 1997), 73–101.

Haury, Emil W. "A Large Pre-Columbian Copper Bell from the Southwest," *American Antiquity* 13:1 (1947), 80–82.

Hoffman, Paul E. "Narváez and Cabeza de Vaca in Florida," in *The Forgotten Centuries: Indians and Europeans in the American South,* 1521–1704, ed. Charles Hudson and Carmen Chaves Tesser (Athens and London: University of Georgia Press, 1994), 50–73.

Hoffman, Paul E. *A New Andalucia and a Way to the Orient: The American Southeast during the Sixteenth Century* (Baton Rouge: Louisiana State University Press, 2004 [1990]).

Hunwick, J. O. "Black Slaves in the Mediterranean World: Introduction to a Neglected Aspect of the African Diaspora," *Slavery and Abolition: A Journal of Comparative Studies,* Special Issue, *The Human Commodity: Perspectives on the Trans-Saharan Slave Trade,* ed. Elizabeth Savage, 13:1 (1992), 5–38.

Konetzke, Richard. *Colección de documentos para la historia de la formación social de Hispano-América:* 1493–1810, 3 vols. (Madrid: Consejo Superior de Investigaciones Científicas, 1953–1962), vol. 1 (1953).

Krieger, Alex. *We Came Naked and Barefoot: The Journey of Cabeza de Vaca across North America,* ed. Marbery H. Krieger (Austin: University of Texas Press, 2002 [1955]).

Kubler, George. *Mexican Architecture of the Sixteenth Century*, 2 vols. (New Haven: Yale University Press, 1948).

Ladd, Edmund J. "Zuni on the Day the Men in Metal Arrived," in *The Coronado Expedition: The 1540–1542 Route across the Southwest*, ed. Richard Flint and Shirley Cushing Flint (Niwot: University Press of Colorado, 1997), 225–233.

Lahon, Didier. "Black African Slaves and Freedmen in Portugal during the Renaissance: Creating a New Pattern of Reality," in *Black Africans in Renaissance Europe*, ed. T. F. Earle and K. J. P. Lowe (Cambridge: Cambridge University Press, 2005), 261–279.

Las Casas, Bartolomé de. *Brevísima relación de la destrucción de las Indias*, ed. Ramón O. Hernández, vol. 10 of *Obras completas* (Madrid: Alianza, 1992 [1552]).

Las Casas, Bartolomé de. *De unico vocationis modo omnium pentium ad veram religiones*, ed. Paulino Castañeda Delgado y Antonio García del Moral, vol. 2 of *Obras completas* (Madrid: Alianza, 1999 [1537]).

Las Casas, Bartolomé de. *Historia de las Indias*, ed. Miguel Angel Medina, Jesús Angel Barreda, and Isacio Pérez Fernández, vols. 3–5 of *Obras completas* (Madrid: Alianza, 1994 [c. 1566]).

Las Casas, Bartolomé de. *Obras completas*, ed. Paulino Castañeda Delgado et al., 14 vols. (Madrid: Alianza, 1992–1998).

Lawson, Edward W. *The Discovery of Florida and Its Discoverer Juan Ponce de León* (Saint Augustine, Florida: Author, 1946).

Leonard, Irving A. "Spanish Ship-Board Reading in the Sixteenth Century," *Hispania* 32:1 (1949), 53–58.

Lesques, Abraham. *The Catalan Atlas of 1375*, ed. Georges Grosjean (Dietikon-Zurich: URS Graf, 1978 [1375]).

Logan, Rayford W. "Estevanico, Negro Discoverer of the Southwest: A Critical Reexamination," *Phylon* 1:4 (1940), 305–314.

López Grigera, Luisa. "Relectura de la *Relación* de Alvar Núñez

Cabeza de Vaca," *Anales de Literatura Hispanoamericana* 28 (1999), 921–932.

López-Portillo y Weber, José. *La conquista de la Nueva Galicia*, Colección de Obras Facsimilares 4 (Guadalajara: Instituto Jalisciense de Antropología e Historia, n.d. [1935]).

Lotarelo y Mori, Emilio, ed. *Entremés de los mirones*, in *Colección de entremeses, loas, bailes, jácaras y mojigangas desde fines del siglo XVI á mediados del XVIII*, vol. 17 (Madrid: Nueva Biblioteca de Autores Españoles, 1911).

Lowe, Kate. "The Stereotyping of Black Africans in Renaissance Europe," in *Black Africans in Renaissance Europe*, ed. T. F. Earle and K. J. P. Lowe (Cambridge: Cambridge University Press, 2005), 17–47.

Lowery, Woodbury. *The Spanish Settlements within the Present Limits of the United States* 1513–1561 (New York: Knickerbocker, 1901).

Mariscal, George. "The Role of Spain in Contemporary Race Theory," *Arizona Journal of Hispanic Cultural Studies* 2 (1998), 7–22.

Luján, Diego Pérez de. *Expedition into New Mexico Made by Antonio de Espejo*, 1582–1583, ed. and trans. George P. Hammond and Agapito Rey (Los Angeles: Quivira Society, 1929).

Martín Casares, Aurelia. *La esclavitud en la Granada del siglo XVI: género, raza y religión* (Granada: Universidad de Granada, 2000).

Maura, Juan Francisco ed., "Introduction," in *Cabeza de Vaca* 2001, 8–72.

Maura, Juan Francisco. "Nuevas interpretaciones sobre las aventuras de Alvar Núñez Cabeza de Vaca, Esteban de Dorantes y Fray Marcos de Niza," *Revista de Estudios Hispanicos. Río Piedras, Puerto Rico* 19:1–2 (2002), 129–154.

McDonald, Dedra S. "Intimacy and Empire: Indian-African Interaction in Spanish Colonial Mexico, 1500–1800," *American Indian Quarterly* 22:1–2 (1998), 134–156.

Mena García, Carmen. "Nuevos datos sobre bastimentos y envases en armadas y flotas de la carrera," *Revista de Indias* 64:231 (2004), 447–484.

Mendieta, Gerónimo de, *Historia eclesiástica Indiana*, ed. Francisco Solano y Pérez-Lila (Madrid: Atlas, 1973).

Mercado, Tomás de. *Suma de Tratados y contratos* (Seville: Hernando Díaz, 1571).

Millares, Agustín, and José I. Mantecón. *Indice y estractos de los Protocolos del Archivo de Notarios de México, DF*, 2 vols. (Mexico: El Colegio de México, 1945).

Montañé Martí, Julio César. *Por los senderos de la Quimera: El Viaje de Fray Marcos de Niza* (Hermosillo: Instituto Sonorense de Cultura, 1995).

Moreno, Isidoro. *La antigua hermandad de los negros de Sevilla: Etnicidad, poder y sociedad en 600 años de historia* (Seville: Universidad de Sevilla and Consejería de Cultura de la Junta de Andalucía, 1997).

Morgado, Alonso de. *Historia de Sevilla* (Seville: Archivo Hispalense 1887 [1587]).

Motolinía, Fray Toribio de Benavente de. *Historia de los Indios de la Nueva España*, ed. Georges Baudot (Madrid: Castalia, 1985 [c. 1541]).

Navagero, Andrea. *Viaje a España . . . etc.*, ed. José María Alonso Gamo (Valencia: Castalia, 1951 [c. 1526]).

Navarro García, Luís. *Sonora y Sinaloa en el siglo XVII* (Sevilla: Escuela de Estudios Hispano-Americanos, 1967).

Newcomb, W.W. *The Indians of Texas from Prehistoric to Modern Times* (Austin: University of Texas Press, 1961).

Obregón, Baltasar. *Historia de los descubrimientos de la Nueva España*, ed. Eva María Bravo (Sevilla: Alfar, 1997 [1584]).

Oliveira Marques, A. H. de, ed., "Uma descriçao de Portugal em 1578–80," Anon., *Portugal Quinhentista (Ensaios)* (Lisbon: Quetzal, 1987), 127–245.

Olsen, Donald N., et al., "Piñon Pines and the Route of Cabeza de Vaca," *Southern Historical Quarterly* 1:2 (1997), 174–186.

Ortega Sagrista, Rafael. "La Cofradía de los Negros en el Jaen del Siglo XVII," *Boletín de Estudios Giennenses* 4:1–2 (1957), 1–10.

Ortiz de Zúñiga, Diego. *Annales eclesiásticos y seculares de la muy noble y muy leal ciudad de Sevilla* (Madrid: Florian Anisson, 1677).

Oviedo y Valdés, Gonzalo Fernández de. *Historia general de las Indias,* ed. Pérez de Tudela, 5 vols. (Madrid: BAE, 1959 [c. 1557]).

Palomero Páramo, Jesús Miguel. "Las procesiones de 'sangre' en Sevilla y Nueva España: A propósito de una nueva pintura mural en la iglesia conventual de Huexotzingo," in *Primeras Jornadas de Andalucía y América,* 2 vols. (Huelva: Instituto de Estudios Onubense, 1981), 1:313–319.

Parry, J. H. *The Audiencia of New Galicia in the Sixteenth Century: A Study in Spanish Colonial Government* (Cambridge: Cambridge University Press, 1948).

Peña Fernández, Francisco. "El *Otro* Héroe: *Los Naufragios* de Cabeza de Vaca como palimpsesto bíblico," *Bandue: Revista de la Sociedad Española de Ciencias de las Religiones* 1 (2007), 179–194.

Peña y Cámara, José de la. "Contribuciones documentas y críticas para una biografía de Gonzalo Fernández de Oviedo," *Revista de Indias* 17 (1957), 603–705.

Pérez-Malaína, Pablo E., *Spain's Men of the Sea: Daily Life on the Indies Fleets in the Sixteenth Century,* trans. Carla Rahn Phillips (Baltimore and London: Johns Hopkins University Press, 1998 [1992]).

Pérez de Ribas, Andrés, *History of the Triumphs of Our Holy Faith amongst the Most Barbarous and Fierce Peoples of the New World,* ed. Daniel T. Reff, Maureen Ahern, and Richard K. Danford (Tucson: University of Arizona Press, 1999 [1645]).

Pérez de Tudela Bueso, Juan. "Estudio preliminar: Vida y escritos de Gonzalo Fernández de Oviedo," (1959), in Oviedo (c. 1557), 1:v–clxxv.

Peter Martyr, *"Archiepiscopo Cusentino,"* in *The Discovery of the New World in the Writings of Peter Martyr of Anghiera,* ed. Ernesto Lundardi, Elisa Magioncalda, Rosanna Mazzacane, trans. Felix Azzola and Luciano F. Farina, in *Nuova Raccolta Colombiana* (English Edition), vol. 10 (Rome: Librería dello Stato, 1992 [1525]), 186–189.

Pike, Ruth. "Crime and Criminals in Sixteenth-Century Spain," *Sixteenth Century Journal* 6:1 (1975), 3–18.

Pike, Ruth. "Sevillian Society in the Sixteenth Century: Slaves and Freedmen," *Hispanic American Historical Review* 47:3 (1967), 344–359.

Ramusio, Giovanni Battista. *Navigationi et Viaggi,* 3 vols. (Venice: Giunti, 1555).

Reff, Daniel T. "The Location of Corazones and Señora: Archaeological Evidence from the Río Sonora Valley, Mexico," in *The Protohistoric Period in the North American Southwest, AD* 1450–1700 (Arizona State University, Anthropological Research Papers, 24), ed. David R. Wilcox and W. Bruce Masse (Tempe: Arizona State University, 1981), 94–112.

Restall, Mathew. *Seven Myths of the Spanish Conquest* (Oxford: Oxford University Press, 2003).

Ricklis, Robert A. *The Karankawa Indians of Texas: An Ecological Study of Cultural Tradition and Change* (Austin: University of Texas Press, 1996).

Rodrigues, Bernardo. *Anais de Arzila: Crónica inédita do séclo XVI,* vol. 1 (Lisbon: Academia das Ciencias de Lisboa, 1915 [c. 1560]).

Rodríguez de Montalvo, Garci. *Amadís de Gaula,* ed. Juan Manuel Cacho Blecua. 2 vols. (Madrid: Cátedra, 1987 [1508]).

Rodríguez Morel, Genaro. *Cartas del Cabildo de la ciudad de Santo*

Domingo en el siglo XVI (Santo Domingo: Patronato de la Ciudad Colonial de Santo Domingo, Amigo del Hogar, 1999).

Rojas Villandrando, Agustín de. *El viaje entretendio*, ed. Jean Pierre Ressot (Madrid: Castalia, 1972 [1604]).

Rosenberger, Bernard, and Hamid Triki. "Famines et épidémies au Maroc aux XVIe et XVIIe siècle," *Hespéris Tamuda* 14 (1973), 109–176.

Rout, Leslie B. *The African Experience in Spanish America: 1502 to the Present Day*, Cambridge Latin American Studies, vol. 23 (Cambridge: Cambridge University Press, 1976).

Ruiz Medrano, Ethelia. *Reshaping New Spain: Government and Private Interests in the Colonial Bureaucracy, 1531–1550*, trans. Julia Constantino and Pauline Marmasse (Boulder: University Press of Colorado, 2006).

Russell, P. E. "La 'poesía negra' de Rodrigo de Reinosa," in *Temas de La Celestina* (Barcelona: Ariel, 1978), 377–406; first published in *Studies in Spanish Literature of the Golden Age Presented to Edward M. Wilson*, ed. R. O. Jones (London: Támesis, 1973), 225–246.

Salazar, Eugenio de. *Cartas* (Madrid, Sociedad de Bibliófilos Españoles, 1866 [c. 1573]).

Sancho de Sopranis, Hipólito. *Las cofradías de morenos en Cadiz* (Madrid: CSIC, 1958).

Santa Cruz, Alonso de. *Crónica del Emperador Carlos V*, ed. Francisco de la Iglesia y Auset (Madrid: Patronato de Húerfanos, 1920 [c. 1540]).

Sauer, Carl O. "The Road to *Cíbola*," *Ibero-Americana* 9 (1934 [1932]), 1–58.

Sauer, Carl O. "The Discovery of New Mexico Reconsidered," *New Mexico Historical Review* 12 (1937), 270–287.

Sauer, Carl O. "The Credibility of the Fray Marcos Account," *New Mexico Historical Review* 16 (1941), 233–246.

Saunders, A. C. de C. M. *A Social History of Black Slaves and Freedmen in Portugal,* 1441–1555 (Cambridge: Cambridge University Press, 1982).

Schaedel, Richard P. "The Karankawa of the Texas Gulf Coast," *Southwestern Journal of Anthropology* 5:2 (1949), 117–137.

Sibley, John. *Message from the President of the United States . . .* (Washington, DC: Way, 1806).

Silva, Alan J. "Conquest, Conversion, and the Hybrid Self in Cabeza de Vaca's *Relación,*" *Post Identity* 2:1 (1999), 123–146.

Smith, Roger C. *Vanguard of Empire: Ships of Exploration in the Age of Columbus* (Oxford and New York: Oxford University Press, 1993).

Sousa, Fray Luiz de [Manoel de Sousa Coutinho]. *Annaes de El Rei Dom Joao Terceiro,* ed. A Herculano (Lisbon: Sociedade Propagadora dos Conhecimentos Uteis, 1844 [c. 1632]).

Tello, Fray Antonio. *Libro segundo de la Crónica miscelanea, en que se trata de la conquista espiritual y temporal de la Santa Provincia de Xalisco etc.,* ed. José Lopez-Portillo y Rojas (Guadalajara: República Literaria, 1891 [c. 1650]).

Teonge, Henry. *The Diary of Henry Teonge,* ed. G. E. Manwaring, (London: Routledge, 1927 [c. 1679]).

Theisen, Gerald. Translation of Oviedo in *The Narrative of Alvar Núñez Cabeza de Vaca* (Barre, MA: Imprint Society, 1972).

Tomson, Robert. *An Englishman and the Mexican Inquisition* 1556–1560, ed. G. R. G. Conway (Mexico: Privately printed, 1927 [c. 1560]).

Torres, Diego de. *Relación del origin y sucesso de los Xarifes* (Seville: Francisco Pérez, 1586).

Troncoso, Franisco P. *Las guerras en las tribus Yaqui y Mayo del Estado de Sonora* (Mexico: Departamento de Estado Mayor, 1905).

Turner, Raymond. "Oviedo's *Claribalte*: The First American Novel," *Romance Notes* 6 (1964), 65–68.

Wagner, Henry R. "Fr. Marcos de Niza," *New Mexico Historical Review* 9 (1934), 184–227.

Weddle, Robert S. *Wilderness Manhunt: The Spanish Search for La Salle* (College Station, Texas A&M University Press, 1999 [1973]).

Woodbury, George, and Edna Woodbury. *Prehistoric Skeletal Remains from the Texas Coast* (Gila Pueblo: Medallion, 1935).

Wright, Richard R. "Negro Companions of the Spanish Explorers," *American Anthropologist* 4:2 (1902), 217–228.

Zorita, Alonso de. *Relación de la Nueva España,* ed. Ethelia Ruiz Medrano, 2 vols. (Mexico: Consejo Nacional para la Cultura y Artes, 1999 [1585]).

Zurara, Gomes Eannes da. *The Chronicle of the Discovery and Conquest of Guinea,* trans. Chares R. Beazley and Edgar Prestage, 2 vols. (London: Hakluyt Society, 1896 [c. 1453]).

NOTES

7   **Major Richard Robert Wright**   Wright 1902; quotation, 225.

17   **In the far northwest of Mexico**   This account of Esteban's life draws on three main sources: Cabeza de Vaca [1542], Oviedo [c. 1557], and Niza [1539]. These sources are not cited hereafter.

18   **"natural gangster"**   This evocative phrase is used by Parry 1948, 22.

18   **he was as audacious, dynamic, and brave as he was bloody**   Arteaga 1932, 481.

19   **Lázaro de Cebreros, to search for the trail to Culiacán**   Pérez de Ribas [1645], 103.

19   **a grim story** Troncoso 1905, 37–38.

19   **Both wore feather headdresses**   Tello [1650], 187.

19   **"Take me to your leader!"**   The exchange is described by Cabeza de Vaca in *Shipwrecks*. I have used both *Shipwrecks* and Oviedo's *History* extensively in my account of the Narváez expedition and the first crossing of America, but references to those books are not supplied in the Notes, because to do so would be very cumbersome indeed.

21   **Cebreros tenderly embraced**   Ibid., 186.

26   **About 500 of these Opata**   Sauer [1932], 20.

26   **mother-of-pearl crucifixes**   These details are from the account of Martín Pérez, a late-sixteenth-century misssionary to the area; see González Rodríguez and Anzures y Bolaños 1996, 188 and 199.

27   **scarred, unnatural eyelids**   González Rodríguez and Anzures y Bolaños 1996, 179–184; Third Anonymous Conquistador's account in López-Portillo y Weber 1935, 363.

27   **Cebreros left them at the Indian town**   Tello [1650], 187.

27   **Díaz must have been**   Ibid., 187. Although Tello is unreliable and may have romanticized this detail about their being old friends, it may equally well have been true: the conquistadors frequently knew each other from Spain or the New World.

28   **whose mother's name was Maldonado**   "Maldonado Bravo" is among the original settlers of Culiacán listed by López-Portillo y Weber 1935, 175 and 361.

28 **embraced and wept together** Cabeza de Vaca claims that Melchior Díaz wept. Weeping, although it was then a more usual form of greeting than it is today in the western world, still suggests a degree of emotion perhaps revealing a previous friendship.

29 **the rivers were full of fish** For the fish: González Rodríguez and Anzures y Bolaños 1996, 181.

29 **"affable" Indians** For the porches: ibid., 202 and 206.

31 **carried in hammocks** Diego de Guzmán met with chiefs in this area who were carried in hammocks; see Portillo 1935, 363.

31 **This peculiar document** Hunwick 1992, 7.

33 **ritual gatherings** González Rodríguez and Anzures y Bolaños 1996, 206–207.

34 **One later missionary** Ibid., 188–189.

38 **trophies of their great journey** This perspicacious observation about the psychology and symbolic importance of their Indian clothes is made by Tello [1650], 187.

38 **Esteban was lodged** Ibid., 189.

39 **somewhere near Guadalajara** There is no documentary evidence that this meeting ever took place. But these three men lived in Guadalajara, and it is difficult to believe that they would have failed to meet the four survivors if, as seems likely, they had the opportunity to do so. There is documentary evidence about what had happened to them, which informs what follows. Also see Goodwin 2007.

39 **Juan Durán later reported** AGI: Patronato 55 N 5 R 4, 1v–2r.

40 **In *Shipwrecks*, Cabeza de Vaca reported** For the theory that information was supplied see Goodwin 2007, 162–164.

40 **A document I came across in Spain** AGI: México 1088 L 3, 251r–251v.

40 **"All they do is sleep, eat, and drink"** *CDI*, 12:24.

40 **This news was enough encouragement** AGI: Patronato 157 N 2 R 4 1556.

40 **They arrived as "defeated soldiers"** AGI: Patronato, 65 N 1 R 4 1562, 18r; AGI: Patronato 55 N 5 R 4, 3v.

44 **The panoply of produce** Zorita [1585], 1:189–190.

47 **Mexico reminded them of the Castle of Bradoid** Gilman 1961, 111.

47 **Bradiod was built on a rocky outcrop** Rodríguez de Montalvo [1508], 1:332–334 [11].

50 **with few exceptions** Two important exceptions are Alegría 1990 and Restall 2003, 55–63.

51 **"I state that I need"** AGI: México 204 N 3.

51 **a claim supported** Gómara [1552], chap. 240.

51 **A document I uncovered at the Archivo de Indias** AGI: Indiferente 423 L 20, ff.528v–529r, 550r–550v, 752v–754r; AGI: Justicia 1173 N 5.

53 **The division was plain to see** The observer is Bernaldo de Valbuena, in Kubler 1948, 75.

53 **In 1555** Tomson and Chilton are quoted ibid.

53 **In 1531** This commentator was Diego Proaño de Hurtado, ibid., 28.

53 **building blocks and paving stones** Ibid., 163.

54 **Calle Tacuba ran straight** The description of Mexico City draws extensively on Cervantes de Salazar [1554], 37–67.

54 **a royal decree was sent to the viceroy** Mendoza to the Crown, 1537, in *CDI*, 2:182

55 **Thomas Gage** Quoted in Bennett 2003, 18–19.

55 **Doña Nufla and Doña Zangamanga** Rojas Villandrando [1604], 217.

57 **Alonso de la Barrera** AGI: Patronato 57 N 4 R 1 1547, 11v.

58 **One missionary describes** Motolinía [c. 1541], 200–201.

58 **During the Corpus Christi processions** Ibid., 213.

58 **But these festivities** García Icazbalceta 1877, xxvii–xxviii.

59 **the true pride of a founding father** Díaz del Castillo [c. 1568], chap. 204.

64 **"We have learned of the ungoverned greed"** "Provisión Real sobre el buen tratamiento de los indios, dada en Granada á 27 de noviembre de 1526," in *CDI*, 1:450–469.

65 **They were sent like convicts** Quoted by Benítez [1953], 88–89.

66 **Juan de Zumárraga's career** Greenleaf 1961, 112.

66 **model their lives on Saint Francis** Cervantes 1991, 4.

67 **Zumárraga was one of the key religious figures** Hartmann 1997, 83–84.

67 **influenced by Bartolomé de Las Casas** The argument for peaceful evangelization was most forcefully laid out in Las Casas [1537].

67 **According to Las Casas** From Las Casas [c. 1566], 3:1865–1881.

69 **On another occasion** Las Casas [1552], 43.

70 **Las Casas was so horrified** Ibid.

70 **That year, in an aggressive letter** 'Parecer al Virey sobre esclavos de rescate y guerra,' in García Icazbalceta 1947, 3:90–94.

71 **the devil's work** Cervantes 1991, 6–7.

71 **Zumárraga could be a brutal Inquisitor** Greenleaf 1978.

72 **Zumárraga now wrote for Mendoza a report** García Icazbalceta 1947, 3:90–94.

73 **"I am quite persuaded"** Quoted ibid.

74 **a "messiah"; the prodigal son** Peña Fernández 2007.

74 **Another scholar points out** López Grigera 1999, 931. See also Adorno 1992, 220–227; Silva 1999.

74 **Antonio de Mendoza treated the four survivors** AGI: Mexico 212 N 45.

75 **Mendoza's viceregal palace** Benítez [1953], 16–17.

81 **"Indeed it seems clear"** Wright 1902, 225.

82 **In the first place, Logan** Logan 1940, 306.

82 **Cleve Hallenbeck, a meteorologist** Hallenbeck 1940.

83 **a Christian brotherhood** Moreno 1997, 45.

84 **In fact, although the Catalan uses *negre*** Lesques [1375], 63; the original is in the Bibliothèque Nationale, Paris, Cat. No. Esp. 30.

85 **In the 1450s an Italian merchant** Cadamosto [1507], 17.

85 **a Portuguese history** Fernandes [c. 1507], f. 338r.

85 **Near the end of the 1500s** Torres 1586, 1–2.

85 **a grant made in 1594** Cortés López 1989, 43.

85   "When it comes to talking about slaves"   Quoted in Castellano 1961, 57.

86   By contrast, most examples   Cortés 1964, 56–57.

86   For there is at least one reference   Ibid., 221 doc 27; 442 doc 1, 397.

86   Many years later, Cortés explained   Cortés 1989, 86.

87   Aurelia Martín has researched   Martín Casares 2000.

87   universally accepted by scholars   Another is Cortés López 1989.

88   as many sub-Saharans were forcibly removed   Hunwick 1992, 5.

94   Gomes Eannes de Zurara's famous *Chronicle of the Discovery and Conquest of Guinea*   The account follows Zurara [c. 1453], 30–61.

95   to quote Zurara's description   Ibid., 80–83].

98   In the late 1510s   The suggestion was first made, so far as I am aware, in Maura 2001.

99   Drought came first, in 1517   Rosenberger and Triki 1973, 117–121. I suggest that plague may have come from Spain because Daza 1523 describes it at Jerez de la Frontera in 1518 and Rosenberger and Triki first note it in Morocco in 1519.

99   A chronicler of the period   Sousa [c. 1632], 59.

99   Bernardo was so struck   For the following, Rodrigues [c. 1560], 326–330.

102   a young Spanish chronicler   This account follows Daza [1523].

106   this has been eloquently disputed   Arens 1978.

107   He had betrayed his humanity   The enslavement of Africans had long been justfied by racist tradition, but the indigenous population of the New World presented a more complex problem for sixteenth-century European philosophers and moralists. There were many who believed the Indians to be in a prelapsarian state, an innocent population untouched by original sin. Others argued that they might be descended from a lost tribe of Israel. Theologians argued that their souls had become the moral responsibility of their European sovereigns.

108   the first religious brotherhood   Sancho de Sopranis 1958, 11.

112   In Seville, in the spring of 1522   Ortiz de Zúñiga 1677, 479.

112   Without documentary proof   Franco Silva 2000, 585; Martín Casares 2005.

112   Moors, Berbers, and Turks   Gestoso y Pérez 1910, 87.

112   marked with an S   Graullera Sanz 1978, 119.

113   Ironically, these steps   Domínguez Ortiz 2003, 10.

113   An Italian ambassador   Navagero [c. 1526], 53–55.

113   A contemporary Sevillian commentator   Morgado [1587], 168–169.

114   The Italian ambassador remembered   Navagero [c. 1526], 57

115   Seville had a long tradition   Rout 1976, 13–15.

115   foreigners compared the city to a game of chess   The observation was made by Alonso Cortés in his "Diálogo en alabanza de Valladolid": see Domínguez Ortiz 2003, 10. Interestingly, the same was said of Lisbon by an anonymous Italian author; see Oliveira Marques 1987, 240.

115   Information from later censuses   A census for 1565 gives a figure of 6,327 slaves out of a total of 85,538 people. In the 1520s the population of the city was probably more than half of 85,000, so there may have been as many as 3,500 or so slaves in total. Pike 1967, 345, sug-

gests that by the second half of the sixteenth century over half the slaves in Seville were black.

115   **many *horros* and their children**   Franco Silva 2000, 574–578.

116   **this black minority was highly visible**   Pike 1967, 353.

116   **One flabbergasted Flemish scholar**   This was Nicolas Clénard, quoted in Fonseca 2005, 114.

116   **According to one theory**   This may be inferred from observations on the Renaissance habit of using black slaves as a means of display, in Lowe 2005, 24–25.

116   **We know from Juan Daza**   Daza [c 1523].

117   **One very reliable modern scholar**   Jean Bodin, quoted in Pike 1967, 45.

118   **Peter Martyr argued that**   Peter Martyr 1525, 186 and 187.

118   **another influential Spanish writer**   Mercado 1571, book 2, chap. 21.

119   **The view of some**   Cortés López 1989, 90.

119   **In 1569 they issued**   Chaves 1904, 37–38.

120   **a widespread distrust of taverns**   Goodwin 2001, 124–138.

120   **one Italian visitor to Lisbon**   Oliveira Marques 1987, 240.

121   **A character in this comedy**   For the following, see Lotarelo y Mori 1911.

122   **the unpleasant banter**   Cortés López 1989, 95.

125   **the first marquis of Tarifa**   Palomero Páramo 1981, 315.

125   ***Los Negritos* rose to prominence**   Moreno 1997, 25–26.

126   **as night fell**   Chaves 1904, 104–105.

127   **festival of Corpus Christi**   Ibid., 223.

127   **eight men with tambourines**   Moreno 1997, 54.

127   **Spaniards had always embraced**   Saunders 1982, 89.

127   **As early as 1451**   Russell 1973, 383.

127   **they dressed as demons**   Moreno 1997, 54.

127   **it was a common practice**   Lowe 2005, 20.

127   **In Lisbon, for example**   Lahon 2005, 265.

127   **In Jaén, in eastern Andalusia**   Ortega Sagrista 1957, 5.

128   **"house of blacks"**   Blumenthal 2005.

128   **calle del Conde Negro**   Ortiz de Zúñiga 1677, 374.

134   **Oviedo was born**   Bolaños 1990, 577, 625–632.

135   **That silence has led**   Peña y Cámara 1957, 603–606.

135   **He met Margarita in Madird**   Oviedo [c. 1557], 1:198.

136   **During that awful birth**   Quoted in Pérez de Tudela 1959, xxxix.

136   **"God lent Margarita"**   Oviedo [c. 1557], 1:198.

138   **the first American novel**   So described in Turner 1964; Arrom 1983, 134, has disagreed.

138   **"first historian of natural history"**   Bolaños 1990, 590; Gerbi 1978, 268.

139   **first European historian of America**   Bolaños 1990, 578.

144   ***Malhado* was the name of a fictional island**   González 1999.

145   **a man's birth was a matter of luck**   Bolaños 1990, 582.

146   **Anne of Aragon sued for divorce**   Maura 2001, 21–22.

148   **the prodigal son**   Peña Fernández 2007.

149   **He is a missionary**   Maura 2001 is one of the few convincing critics of this sanctified image.

150   **Charles V ordered his officials**   AGI: Indiferente 1962 L 5, ff.273v–267v

151   **a Portuguese nobleman commented**   Elvas 1965 38–39.

153   **The document which seemed to me to prove**   AGI: Méxcio 212 N 45.

153   **some kind of report**   Dorantes de Carranza [1604].

154   **"erstwhile palace of the Moorish kings"**   Navagero [c. 1526], 55

160   **a Spanish dictionary that was published in 1611**   Covarrubias 1611, 947.

164   **Alonso de Santa Cruz had reported**   Santa Cruz [c.1540], 479.

164   **some reason, which Granado does not record**   AGI: Mexico 212 N 45.

164   **innocent passengers**   *CDI* 2:189–190.

165   **Mendoza also stated in his letter**   Ibid., 206–207.

165   **Mendoza later reported**   Hammond and Rey 1940, 2:51–52.

165   **the whole business had failed**   Ibid.

165   **Pedro de Benevides knew only too well**   AGI: Mexico 212 N 45.

166   **we can see the hand of the powerful Archbishop**   García Icazbalceta 1947, 3:90–94.

166   **The *Dorantes Report* contains**   AGI: Mexico 212 N 45.

166   **we know that Dorantes was still in Mexico**   *CDI*, 15:375

166   **we also know**   AGI: Mexico 212 N 45.

166   **a royal decree dated December 1539**   AGI: Lima 565 L 3, f.161v.

167   **between February and November of 1541**   For this argument in detail, see Goodwin 2007, 162–164. Castillo had returned by November, when he acted as a witness in Puebla; see Boyd-Bowman 1988, 1:38.

174   **Oviedo was clear about**   Oviedo [c. 1557], 2:402.

175   **given permission to take four black slaves**   AGI: Indiferente 421 L 12, 50v–50r and 71v–72r.

176   **They were headed for the Atlantic port**   AGI: Patronato 55 N 5 R 4, 1v.

176   **The contract may have stated**   These instructions relating to the quality of the wine for a late-sixteenth-century voyage are quoted in Mena García 2004, 451.

176   **Many passengers brought their own food**   For rations supplied to Pedro Menéndez de Avilés's expedition to Florida in 1568, see Pérez-Malaína 1988, 141.

177   **One English traveler recalled**   Teonge [c. 1679], 233.

177   **coffinlike cabins**   Pérez-Malaína 1988, 138.

179   **the young Alonso del Castillo Maldonado**   *CDI* 41:272–273.

179   **Eugenio de Salazar recited**   Salazar [c. 1573], 35–36.

179   **Zumárraga complained**   Cuevas and Genero 1975, 55–56.

179   **a friar who accompanied Bartolomé de las Casas**   Pérez-Malaína 1988, 129 and 136, quotes Tomás de la Torre, "Diario del viaje de Salamanca a Ciudad Real [Chiapas], 1544–1545."

180 **Salazar recalled how he squeezed** This description is closely based on Salazar [c. 1573], 36–38.

181 **working the pumps** Pérez-Malaína 1988, 71.

182 **Salazar now turned his sarcasm** Salazar [c. 1573], 43–46.

183 **the best-selling novel *Amadis of Gaul*** Leonard 1949.

183 **his moral censure** Pérez-Malaína 1988, 168.

183 **A boy called Pedro Merino** Ibid., 172–174.

185 **on one seventeenth-century crossing** Gutiérrez de Medina 1947, 21–22.

185 **possibly written by Castillo** The argument that Castillo contributed in some way to the final chapter of *Shipwrecks* is laid out in Goodwin 2007, 162–164.

189 **A century and a half later** Weddle 1973, 2.

190 **The intrepid English traveler Robert Tomson** Tomson [c. 1560], 6.

191 **Interest in the Spanish Caribbean islands** Adorno and Pautz 1999, 2:41; this three-volume edition and study of *Shipwrecks* is an essential, almost encyclopedic reference work, a comprehensive survey of sources that has informed much of my research.

191 **One deserter, Francisco Diaz** AGI: Patronato 157 N 2 R 4.

191 **Santo Domingo was also a terrifyingly expensive place** Rodríguez Morel 1999, 72, 76, 58, 60, and 72–73.

193 **Antonio de Aguayo** Goodwin 2007, 154–157; AGI: Patronato 65 N 1 R 4.

193 **Pedro de Valdivieso** Boyd-Bowman 1985, No. 2565.

194 **It is Oviedo who tells us the sorry tale** Oviedo [c. 1557], 4:322–330.

195 **Paul Hoffman** Hoffman 1990, 78 n.43.

196 **one of his cronies, a local official** Vasco Porcallo was a regidor; AGI: Indiferente 421 L 11, f77v.

197 **"A pilot is to a ship"** Quoted in Smith 1993, 138.

198 **Velasco wrote** Quoted in Lawson 1946, 22 n.56.

199 **Paul Hoffman offers** Hoffman 1994, 56.

207 **As the Karankawa had always done** Ricklis 1996, 107–108.

208 **constantly referred to by travelers** For details about the Karankawa: ibid., 9–14. Archaeological finds show Karankawa skeletal remains to be especially large, according to Woodbury and Woodbury 1935, 43.

211 **"maze of islands"** Ricklis 1996, 11.

223 **Jéronimo de Alanís** AGI: Indiferente 419 L 4, 106r–107v and 138v; 420 L 8, 193r–v.

224 **It is far from clear** Oviedo says they left both Cabeza de Vaca and León behind, but *Shipwrecks* says they took León with them.

225 **Friar Juan Suárez** AGI: Indiferente 421 L 13, 60r.

232 **Oviedo's *History* reports** *E allí se quedaron en aquel rancho estos dos hidalgos e un negro que les pareció que bastaba para lo que los indios los querían que era para que les acarreasen a cuestas leña e agua e servirse dellos como de esclavos.* Quoted from the Colombina manuscript.

232 **that representation of the situation** Theisen 1972.

236 **the survivors were forced to do women's work** See Armas Wilson 1993.

236 **common among the Karankawa** Schaedel 1949, 125.

236 **This is perhaps the earliest description** This summary of the berdache is derived from Callender and Kochems 1983.

238 **In the summer of 1530** The published edition of Oviedo commonly used (Pérez Tudela, ed., 1959) is confusing, but the manuscript in the Colombina Library makes this date clear.

243 **Esteban and Castillo sucked and chewed** I have supplemented the account in *Shipwrecks* with a description by the nineteenth-century frontierswoman Alice Cooper; see Gatschet 1891.

244 **inspired by a decree** Konetzke 1953, 169. The decree was issued in 1535 and again in 1538.

247 **a black witch doctor called Lucas Olola** For the following account of Olola see Aguirre Beltrán 1963, 68–69.

247 **Guatesco Indians** Newcomb 1961, 29

248 **Esteban had fallen out** Pedro de Castañeda's narrative; see Flint and Flint 2005, 439.

261 **John Sibley** Sibley 1806, 72–73.

268 **the lone voice of Henry Dobyns** Dobyns 1966, 1983.

269 **"It may be no exaggeration"** Dobyns 1983, 102.

274 **Cabeza de Vaca might as well have been describing** Maura 2001, 181 n.85.

274 **Alex Krieger, a keen sleuth of the route** Krieger [1955], 67–68.

274 **Scientists have since found** Olsen et al. 1997.

275 **discovery of a similar bell** Haury 1947, 80–81.

275 **Casas Grandes site** Epstein 1991, 478.

276 **"wings like a fluke"** Sibley 1806, 73.

283 **They danced to the rhythm of haunting voices** Luján [1583], 67.

283 **"skins attached to a vessel in the form of a tambourine"** Hernán Gallegos's account, in Hammond and Rey 1966, 78.

283 **The naked dancers** Obregón [1584], 235.

283 **To the Spaniards this dance** Hammond and Rey 1966, 78.

284 **It would have been unwise** Driver 1961, 111–113, map 13.

285 **The indians warmly welcomed** Luján [1583], 62; Hammond and Rey 1966, 77.

285 **The chroniclers of both these later forays** Luján [1583], 60; Hammond and Rey 1966, 75–77.

286 **pits like a prickly pear** Luján [1583], 65 n.45.

286 **"They also grew cotton"** Hammond and Rey 1966, 78–79.

287 **They passed through Arizpe** Obregón [1584], 159.

287 **north of modern Baviácora** Reff 1981, 104.

287 **described by Las Casas** Las Casas [1566], 3:1160.

288 **Deer, hare, wolves** Ibid.

289 **"Nearby," he said** Tello [1650], 184.

289 **One Spanish conquistador** Alarcón and Castañeda, in Flint and Flint 2005, 216, 439.

297 **"Seven Portuguese Bishops"** See, e.g., Lowery 1901, 256.

297　a story told to Nuño de Guzmán　Flint and Flint 2005, 386.

297　In 1527, as Narváez prepared　Luís de Cárdenas; see Sauer 1937, 271.

298　In 1537, as Mendoza　Ruíz Medrano 2006, 117–120.

299　Mendoza was firm　The words are reported in Francisco Vázquez de Coronado's testimony to Tello de Sandoval's judicial enquiry into the running of New Spain; see Aiton and Rey 1937, 311.

299　In February of the following year　AGI: Patronato 184 R 27.

300　Castillo was very young　*CDI*, 41:272–273.

300　Mendoza arranged for Castillo to marry　AGI: Patronato 275 R 39.

300　Puebla de los Angeles, where his wife was a landower　He is referred to in a number of legal documents; see Boyd-Bowman 1988, 1:19; as a landowner, see AGI: Patronato 278 N 2 R 230.

300　Mendoza wrote to Charles　V *CDI*, 2:206–207.

301　Obregón reported that he then agreed　Obregón [1584], 48.

302　One would-be biographer　Nakayama (1975, 25), who is quoted by Montañé Martí 1995, 19 n.13.

303　Equally, we would do well　These two reports of Esteban as a womanizer (and sexual predator) cannot be dismissed, but should be treated with particular caution. One is Castañeda's comment, mentioned above; see Flint and Flint 2005, 439. The other is in Coronado's account of his arrival at "Cíbola," written to Mendoza under stressful circumstances in 1540; see Flint and Flint 2005, 270.

303　pioneer African-Americans　See, e.g., Aguirre Beltrán [1946], 15.

304　Astonishingly it is an argument　Ibid., 157.

305　A royal decree explained　AGI: Indiferente 427 R 30, f.248.

305　enslaved Africans also managed to own property　Konetzke 1953, 502.

305　Africans frequently appear in legal documents　Gibson 1964, 279.

305　In 1541, the Spanish authorities　Konetzke 1953, 206.

305　The Spanish government reacted　Gibson 1964, 147.

306　In practice, attempts to separate　The Crown time and again reiterated the ban against blacks living among Indians, each new repetition of the rules indicating the extent to which they were being infringed. Konetzke 1953 records new promulgations of the laws in 1541, 1551, 1554, 1567, 1578, 1580, and 1587 (pp. 213, 291, 297, 321, 422, 513, 527, 566, 584).

307　The Crown had decreed　Ibid., 163.

307　In such a climate　*CDI*, 2:198–199.

308　Years later, in 1612　For the account of events in 1612, see Chimalpahin Cuauhtlehuanitzin 1998, 1:521–525.

310　An angry colonial official　The official was Lope de Samaniego; see AGI: México 95 N 39.

311　a royal decree from Charles V　Cited in Flint and Flint 2005, 45.

311　a good navigator and geographer　For Marcos's report and instructions I have followed AGI: Patronato 20 N 5 R 10; see also Flint and Flint 2005, 65–77.

311 On April 4, 1537   Cuevas and Genero 1975, 83–84.

312 **how deeply Zumárraga trusted Marcos**   For the correspondence, see García Icazbalceta 1947, 3:263–265.

313 **from the pulpit and in the barbers' shops**   *CDI* 5:397; Flint and Flint 2005, 104.

314 **"home to elephants and camels"**   García Icazbalceta 1886, 1:194–195.

314 **Even the shrewd Zumárraga was taken in**   García Icazbalceta 1947, 3:186.

317 **In late 1538, Coronado set out**   AGI: Patronato 20 N 5 R 10.

319 **their own estates and private purposes**   Coronado to Crown, 1538; in Flint and Flint 2005, 27.

319 **"Exhort and encourage"**   AGI: Patronato 20 N 5 R 10.

321 **The landowners proposed moving Compostela**   Coronado to Crown, 1538; in Flint and Flint 2005, 29.

321 **steep, craggy roads**   The description is based on a an early-seventeenth-century account by Marcos de Tapia, official of the Audiencia of New Galicia, quoted in Navarro García 1967, 41.

322 **As Coronado reported to his sovereign**   The following draws on Coronado's letter to the king, in Flint and Flint 2005, 42–44.

324 **After twenty days of discussion**   For this and the following: Flint and Flint 2005, 43 (Coronado) and 50 (Mendoza).

325 **a quite bizarre letter**   The letter was published by Giovanni Battista Ramusio [1555], an Italian who collected and edited travel accounts; see Flint and Flint 2005, 31–36. Ramusio's versions of texts were sometimes embellished and are not completely reliable. Flint and Flint (32) argue that the first of these was fabricated. On the other hand, if Coronado and Mendoza were engaged in propaganda, that would also explain why the first account is so exaggerated.

327 **the two Franciscans set out**   Sauer [1932], 3.

328 **found indecent and immoral**   Flint and Flint 2005, 439.

328 **the rough-and-ready world**   The description is based on "Segunda relación anónima," in García Icazbalceta 1866, 2:296.

330 **the accusation that he was a "liar"**   Bloom 1941 asks this question: "Was Fray Marcos a Liar?"

330 **Baltazar Obregón summed up**   Obregón [1584], 51.

331 **As Sauer put it**   Sauer [1932]; quotation from Sauer 1937, 279.

331 **The same conclusion was reached by Cleve Hallenbeck**   Hallenbeck 1949.

331 **Lansing Bloom noticed**   Bloom 1940, 1941.

331 **Sauer then retorted**   Sauer 1941.

332 **every historian who attempts**   However, it is relatively easy to show that Vacapa probably lay much farther south than the defense would like us to believe. Marcos said that he left the Spanish colony of San Miguel de Culiacán on March 7 and arrived at Vacapa on March 21, a journey of fourteen days, of which three were spent at Petatlán. The various sources state that the distance from San Miguel, which was on the San Lorenzo River, to Petatlán was between thirty and fifty leagues. None of the sources suggest that

it was more than fifty. Marcos tells us that it was twenty-five or thirty leagues from Petatlán to Vacapa. Vacapa was therefore eighty leagues from San Miguel. There were about three leagues to a mile, but precise accuracy is not important, because Marcos had no way of accurately measuring the distances and it is not entirely clear whether the figures he gives refer to road miles or estimates of distances "as the crow flies." Vacapa should be about 240 miles north of San Miguel. At that point, not far from the town of El Fuerte, on the Fuerte River, where the spectacular Copper Canyon railway line begins its tortuous ascent into the Sierra Madre, is a place called Vaca, which Carl Otwin Sauer long ago identified as Vacapa.

332 **Another investigator** Wagner 1934, 184.

333 **Coronado found himself forced** Coronado to Mendoza, August 3, 1540, in Flint and Flint 2005, 252.

333 **Sauer argued that Mendoza ensured** Sauer 1937, 270, 287.

333 **There are scholars who disagree** Flint and Flint 2005, 60.

336 **some mildly alcoholic drink made from mesquite** As reported in "Segunda relación anónima," García Icazbalceta 1886, 2:304.

336 **The chronicler Pedro de Castañeda** Flint and Flint 2005, 439.

349 **Alarcón's account of his exploration** "Narrative of Alarcón's Voyage, 1540," in Flint and Flint 2005, 185–205.

351 **Years later, Castañeda reported that** Flint and Flint 2005, 439–440.

351 **As early as April 1540, after Coronado** Mendoza to Crown, 1540, in Flint and Flint 2005, 241.

354 **"Should I die"** Girón [1530]; translated by the author.

357 **Coronado reported to Mendoza** Coronado to Mendoza, 1540, in Flint and Flint 2005, 270.

358 **During his peace negotiations, Coronado heard that** Flint 2002, 180.

361 **A group of elder Zunis told Cushing** Cushing 1920, 362–363.

361 **"Like the teller of Indian tales"** Cushing 1885.

363 **Cushing told his Boston audience** See Green's comments in Cushing 1885, 172.

368 **some have implied that in this way** McDonald 1998.

373 **"when Esteban reached the Mayo River"** Sauer [1932], 32 n. 37.

INDEX